Springer Theses

Recognizing Outstanding Ph.D. Research

Aims and Scope

The series "Springer Theses" brings together a selection of the very best Ph.D. theses from around the world and across the physical sciences. Nominated and endorsed by two recognized specialists, each published volume has been selected for its scientific excellence and the high impact of its contents for the pertinent field of research. For greater accessibility to non-specialists, the published versions include an extended introduction, as well as a foreword by the student's supervisor explaining the special relevance of the work for the field. As a whole, the series will provide a valuable resource both for newcomers to the research fields described, and for other scientists seeking detailed background information on special questions. Finally, it provides an accredited documentation of the valuable contributions made by today's younger generation of scientists.

Theses are accepted into the series by invited nomination only and must fulfill all of the following criteria

- They must be written in good English.
- The topic should fall within the confines of Chemistry, Physics, Earth Sciences and related interdisciplinary fields such as Materials, Nanoscience, Chemical Engineering, Complex Systems and Biophysics.
- The work reported in the thesis must represent a significant scientific advance.
- If the thesis includes previously published material, permission to reproduce this must be gained from the respective copyright holder.
- They must have been examined and passed during the 12 months prior to nomination.
- Each thesis should include a foreword by the supervisor outlining the significance of its content.
- The theses should have a clearly defined structure including an introduction accessible to scientists not expert in that particular field.

Yoav Sagi

Collisional Narrowing and Dynamical Decoupling in a Dense Ensemble of Cold Atoms

Doctoral Thesis accepted by
the Weizmann Institute of Science,
Rehovot, Israel

 Springer

Author
Dr. Yoav Sagi
Weizmann Institute of Science
Rehovot
Israel

Supervisor
Prof. Dr. Nir Davidson
Weizmann Institute of Science
Rehovot
Israel

ISSN 2190-5053   ISSN 2190-5061   (electronic)
ISBN 978-3-642-44566-8   ISBN 978-3-642-29605-5 (eBook)
DOI 10.1007/978-3-642-29605-5
Springer Heidelberg New York Dordrecht London

Printed on acid-free paper

Springer is part of Springer Science+Business Media (www.springer.com)

Supervisor's Foreword

Ultra-cold atomic ensembles have played a key role in advancing many research fields in recent years, including precision measurement, quantum many-body physics and quantum information processing. In the latter, atomic ensembles have been used in the implementation of several cornerstones such as long-lived quantum memories, non-classical photon sources and quantum repeaters for long range quantum networks. This Ph.D. work of Dr. Yoav Sagi studies the coherence properties of trapped atomic ensembles at high densities, which are essential to many of the aforementioned applications. The study focuses on how interparticle interactions modify the ensemble coherence dynamics, and whether it is possible to extend the coherence time by means of external control.

The first question addressed is how decoherence is modified due to elastic collisions. Information is stored in the ensemble by encoding it into the coherence between two internal states of the atom. These states then couple differently to the external trapping potential (e.g. laser field) and this in turn leads to broadening of the original narrow transition. Intriguingly, time-dependent fluctuations, which are frequent due to the high density, narrow this broadened transition and prolong the coherence time—a phenomenon first observed in NMR where it is named "motional narrowing". The thesis reports on experiments revealing the analogous effect in optically trapped ^{87}Rb atoms. A beautiful universal dependency of the emergent new coherent timescale on the atomic phase space density is proven to exist. An important related question is how this effect depends on the microscopic physical model of the fluctuations. To answer this question, a discrete fluctuation model is considered, for which a closed-form formula for the spectrum exists. From the model one learns that the motional-narrowed spectrum is sensitive to the specific collision model, i.e. to whether the collisions are "hard" billiard ball like or "soft" forward scattering processes. Experimentally, the model is shown to correctly describe the spectrum of optically trapped atoms without fitting parameters.

The work also addresses the question of whether the coherence time can be extended by applying external control fields. It is well known that fluctuations at low frequencies can be overcome by a single population inverting pulse—the celebrated coherence echo technique. As the collision rate increases a single pulse can no longer achieve this since the average frequency of each atom before and after this pulse is different. However, dynamical decoupling theories generalize this idea to multi-pulse sequences and enable the suppression of noise at higher frequencies. Applying these ideas to atomic ensembles, a 20-fold increase of the coherence time is demonstrated when a sequence with more than 200 pi-pulses is applied. A full characterization of this process is accomplished using quantum process tomography, establishing that a dense ensemble can indeed be used as a quantum memory with coherence times exceeding 3 s.

Finally, the experiments presented here were performed with a new set-up designed and built from scratch by Dr. Sagi. The new all-optical apparatus achieves quantum degeneracy in about 10 s. This comparably short time enables acquisition of large data sets, which of course improve the overall signal to noise. Many important details regarding the construction of the new set-up appear in a separate chapter in this thesis. Although the main scientific achievements reported in this thesis were published in several journal papers, I do believe that a well written thesis has added value: it tells the story from beginning to end without constraints of space or popularity trends. A good thesis makes it easier for a non-professional to become familiar with the research subject, and for a professional to delve into the details sometimes omitted in short journal letters. The present thesis is a good example of this added value, and I am sure the reader will find its study highly rewarding.

Rehovot, Israel, January 2012   Dr. Nir Davidson

Acknowledgments

The last 5 years of my life were the best. I was lucky to work in the wonderful and supportive environment of the Weizmann Institute of Science. The institute is immersed in a sea of grass and flowers which were a perfect surrounding for the games of my children that were also born in this period. I am deeply grateful to many people who helped and supported me along the way:

I am indebted in so many ways to Professor Nir Davidson, my Ph.D. adviser. Nir is a role model; he is a gifted experimenter whose advices in the lab are true time-savers. On the theory side, he delves deep in order to understand the details of any derivation, and to gain intuition for the result. As an adviser, he keeps the balance between guidance and freedom. He knows how to contribute ideas without shading those of his students. He is always available for a discussion; the amount of time he dedicates to his students is truly admirable. In the long and frustrating days of the lab construction, he always projected optimism and assurance. Nir is the first and most thorough referee of any work we do; I feel privileged to have been his student.

To Dr. Amnon Fisher, who led me in my first steps as an experimental physicist and revealed to me the wonders of the lab. I am grateful to Amnon not only for teaching me the fine details of every instrument, circuit or element in the lab, but also for teaching me that the most important merits in the lab are patience, persistence and dedication. Amnon is much more than a mentor—he is a friend.

To Professor Amiram Ron, my M.Sc. adviser, who taught me how to ask the right questions and not to take for granted any assumption along the way. Amiram also taught me not to be afraid of long and very detailed theoretical calculations. I will always admire Amiram for his curiosity in physics that does not decrease even after many decades of research.

I am also grateful for the guidance I received from my Ph.D. committee: Professor David Tannor and Dr. Roee Ozeri.

More than anything, the discussions with my fellow students were the stimulus for the development of the ideas in this thesis. Ido Almog is my friend and partner. His help in the lab construction and contributions to this work are invaluable. I greatly benefited from the many discussions we had, and I feel fully confident in leaving the lab in his skilled hands. My friend Dr. Rami Pugatch was another source of inspiration and support. Rami is a remarkably knowledgeable physicist and does not cease to surprise me with his creativity and originality. It was so pleasant to stroll in the corridors of the department thanks to all my other friends: Shlomi Kotler, Ori Katz, Oren Raz, Dr. Yaron Bromberg, Dr. Yoav Lahini, Yinnon Glickman, Nitzan Akerman, Yonatan Dallal, Dr. Haim Suchowski and others. This gang was truly a unique bunch of smart and nice people that made the department of physics of complex systems such a special place. As the saying from the Talmud goes (Babylonian Talmud, Taanit 7a): "Much have I learned from my masters, and more from any colleagues than from my masters". Special thanks also to the technical and administrative staff of the department: Dr. Rostyslav Baron, Yeruham Shimoni, Gershon Elazar, Guy Han, Israel Gonen, Perla Zalcberg, Rachel Goldman, Malka Paz and Yossi Drier.

This work would have never been accomplished without the support of my family. My beloved parents, Rachel and Yigal Sagi, always gave me the best education in-house and outside. Along the way they were always confident in my abilities. They were loving and encouraging, and the true happiness each achievement of mine gave them strengthened me when my spirit was down. My sisters Leora and Ariela were also supportive and caring, and shared the burden of emotions in good times but also in hard times. Thanks to Lili and Bruno Segal, my in-laws, for all their help. The fruits I reap are all thanks to the generation of my grandparents, Hasia and Jacob Goldenberg and Moshe and Esther Sagi. This generation devoted their life to the resurrection of a sovereign Jewish nation. My grandmother Hasia was expelled from high school during the Second World War because of her Jewish religion, and therefore was never able to fulfill her dream to become a doctor. For her, my achievement was a partial compensation. Though she lived to see the completion of this work, regretfully she is not here to read these lines.

Finally, there are no words to express my gratitude to my wife, Eva. Without you, my love, none of this would have happened. Your absolute belief in me and love without limits gave me the strength to continue even when it was hard. Besides your own Ph.D. in Physics, you have made during this period the most remarkable thing for me—you have made me a father. Forever I will be grateful for all you are giving to me.

Contents

Chapter 1
Introduction

In current day computer and communication technology, information is represented in bits. Computation is performed by applying a series of logical gates to the bits, and in between storing their value in a memory. Quantum computation takes these ideas to the the quantum world. Loosely speaking, using the superposition concept, many computations can be done in parallel thus achieving an exponential speed up. Quantum bits (qubits) replace the classical bit, with the important feature that they are capable of being in a coherent superposition of the two underlying logical states. Other counterparts of classical computation span from quantum memories to quantum processors.

From a practical point of view, a qubit can be implemented as any system with two energy levels. We shall refer to it in this work as a two-level system (TLS). In most isolated quantum systems there are many discrete levels, but we will restrict transitions and measurements to only two of them, and treat it as an effective TLS. Many physical realizations of TLS exist, including photons, ions, quantum dots, Josephson junction loops and atoms. Each of these systems has advantages and disadvantages, and it seems that ultimately a quantum network will combine several of them to benefit from their very different properties [1].

Two of the most used physical qubits are photons and neutral atoms. Photons are easy to manipulate and produce. They interact weakly with their environment and therefore can remain coherent for long travel distances. This last advantage is also their disadvantage: interaction between photons is usually very small, making the implementation of an all optical two-qubit gate very difficult. Atoms, on the other hand, are easy to keep in one place, and can interact strongly with both other atoms and external electromagnetic fields. It is therefore sensible to use atoms as "stationary qubits" for storage and manipulation and use photons as "flying qubits" that interchange the information between separated sites.

One of the controlled ways of interaction between atoms and photons which is widely used is the electro-magnetically induced transparency (EIT) [2]. In this scheme atoms with a lambda-shape energy structure interact with two light fields called "pump" and "probe". The pump is usually much stronger than the probe, and

Y. Sagi, *Collisional Narrowing and Dynamical Decoupling in a Dense Ensemble of Cold Atoms*, Springer Theses, DOI: 10.1007/978-3-642-29605-5_1,

is used to control the interaction strength between the probe and the atomic ensemble. Closing the pump light while the probe is propagating in the atomic ensemble leads to a conversion of the photonic excitation into the coherence between the two low lying states of the atoms. This is sometimes called "storage of light", although only the coherence which was carried by the light is actually stored in the ensemble. The beauty of this conversion process is that it is reversible, which makes the atomic ensemble a true memory.

In order to increase the efficiency of the storage and retrieval processes, it is desired to work with atomic ensembles with high optical depth [3, 4]. This is because the coupling of the atoms to the external electromagnetic field scales as the square root of the number of atoms in a volume where the light intensity is approximately uniform. Working at high optical depth, however, usually implies that the atomic density is high and the rate of inter-particle collisions is large compared to the storage time. One of the goals of this work is to understand how this fact changes the time dynamics of a stored coherence. The question we address is how rapid velocity changing elastic collisions change the decay function of the ensemble coherence, and what is the asymptotic coherence time in such a scenario.

The ensemble I consider in this thesis is that of cold atoms trapped in a conservative potential. In practice, the atoms are cooled and trapped by lasers, and are very well isolated from their surrounding. When trapped in an optical dipole potential, the atoms can be kept for many seconds without scattering photons or interacting with other atoms coming from the walls of the vacuum chamber. This, together with the excellent controllability one has over the experimental parameters, makes this system ideal for studying the effect elastic collisions have on the ensemble coherence. Another point which simplifies things in a cold ensemble is the fact that s-wave scattering is the dominant collision process.

From the point of view of a particular atom, other atoms can be regarded as the environment, and collisions with them can be treated as the coupling to a bath. Also, since the phase space density is low enough, the motion of each atom is to very good approximation classical. The theoretical framework I adopt in this thesis, therefore, will be that of an effective two level system in contact with a Lorentzian reservoir. The experimental results confirm that this is indeed a good approximation.

Using this framework, we find that elastic collisions lead to a phenomenon we call "collisional narrowing", in analogy to the celebrated motional narrowing effect known for many years in the field of nuclear magnetic resonance (NMR). The way this phenomenon manifests itself in the case of cold atomic ensembles is in the increase of the coherence time as the collision rate increases. Though similar to NMR, the effect of collisional narrowing bears some unique features. The newly emerging prolonged coherence time is found to scale universally with the atomic phase space density. This means that in an ensemble dominated by the collisional narrowing effect, adiabatic changes in the trapping potential will not affect the coherence time. A more elaborate theoretical study reveals how the exact shape of coherence decay depends on the distribution of atomic detunings or on the microscopic collision process.

Since the concept of coherent superposition lies in the heart of quantum computation, it is obvious that decoherence is a major obstacle to its realization. Several techniques have been developed throughout the years to cope with this problem, ranging from quantum error correction, decoherence free subspaces and dynamical decoupling. The latter refers to the use of external fields in order to average out the negative effect of the surrounding. Though the positive effect of collisions is to enhance the coherence time, they also introduce a negative outcome; they render the conventional Hahn echo technique useless, since on average the detuning of each atom before and after the echo pulse is different. Another goal of this work is to explore the possibility to extend the ensemble coherence by dynamical decoupling — the generalization of the coherent echo technique to multiple pulses [5–8].

1.1 Outline

Chronologically, the first three years of the Ph.D. work were devoted to the design and construction of the new experimental apparatus. In the experiment, ^{87}Rb atoms are trapped in a dipole potential created by a far-off-resonance laser, and can be cooled down to quantum degeneracy (Bose-Einstein condensation). A summary of the new setup is given in the beginning of Chap. 2, and more details regarding various aspects of the new apparatus are given in following sections.

The theoretical framework is presented in Chap. 3 where I introduce an effective single spin Hamiltonian and define what is the ensemble coherence. I also explain why this observable is indeed measured in a time-domain Ramsey experiment. In Chap. 4 I explore the asymptotic effect of elastic collisions on the ensemble coherence. I show that the coherence time is linearly increasing for increasing collision rate. In what follows I demonstrate the universal scaling of the prolonged coherence time with the atomic phase space density.

These findings are independent of the exact shape of the detuning distribution as long as the observation time is large compared to the mean time between collisions and as long as the detuning distribution has a finite second moment. However, in Chap. 5 I study to much more detail the exact shape of the spectrum and its dependence of the collision model and the detuning distribution. For this purpose I present a specific discrete fluctuation model for which it is possible to derive a closed form formula for the spectrum. This formula holds true for any detuning distribution. I first use this formula to study the case where the detuning distribution is Gaussian. It is shown that even in this case the spectrum deviates from known results for a continuous Gaussian fluctuations. This is especially surprising since for both models the detuning distributions and correlation functions are the same, and the only difference originates from the "softness" of the collisions. An important question which arise in this context is whether the discrete model actually describes the true physics of trapped cold atomic ensemble. To answer this question 3D Monte Carlo simulations and experiments are presented, both providing an affirmative answer.

The motional narrowing effect appears in many physical systems, but in Chap. 6 I study whether this effect can be actually reversed. In other words, the question raised is whether it is possible for fluctuations to broaden the spectrum. I show that the answer is yes, and that the condition for this to happen is that the initial detuning distribution will have heavy-tails with a diverging first moment. Physically, heavy-tails can be maintained only up to some cutoff, which translates into a broadening effect that persists up to some characteristic collision rate. I also explain in this Chapter the analogy between the problems of the ensemble coherence and spatial diffusion, and use this to propose an experiment observing motional narrowing in the latter.

Finally, in Chap. 7 the possibility to extend the coherence time using dynamical decoupling with external control pulses is explored. I present experimental results showing a 20-fold increase of the coherence time when a dynamical decoupling sequence with more than 200 pi-pulses is applied. Using quantum process tomography, it is demonstrated that a dense ensemble with an optical depth of 230 can be used as an atomic memory with coherence times exceeding 3 s. Measurements of the coherence time as a function of the decoupling scheme pulse rate show quadratic scaling, and I explain the origin for this behavior.

References

1. Duan LM, Lukin MD, Cirac JI, Zoller P (2001) Nature 414:413
2. Lukin MD (2003) Rev Mod Phys 75:457
3. Harris SE, Hau LV (1999) Phys Rev Lett 82:4611
4. Gorshkov AV, André A, Fleischhauer M, Sorensen AS, Lukin MD (2007) Phys Rev Lett 98:123601
5. Viola L, Lloyd S, Knill E (1999) Phys Rev Lett 83:4888
6. Kofman AG, Kurizki G (2001) Phys Rev Lett 87:270405
7. Uhrig GS (2007) Phys Rev Lett 98:100504
8. Cywinski L, Lutchyn RM, Nave CP, Sarma SD (2008) Phys Rev B 77:174509

Chapter 2
The Experimental Setup

The experiments described in this work were performed in a new experimental setup designed and built during the PhD. I first give a brief description of the new apparatus which is necessary for the understanding of the experiments. Following, I elaborate on the various subsystems of the new setup. In particular, I describe the sequence which leads to Bose-Einstein condensation.

2.1 Brief Description of the New Apparatus

In the experiment, cold ^{87}Rb atoms are trapped in a far-off-resonance laser (see Fig. 2.1). The two relevant internal states are $|1\rangle = |F = 1; m_f = -1\rangle$ and $|2\rangle = |F = 2; m_f = 1\rangle$ in the $5^2 S_{1/2}$ manifold, which are, to first order, Zeeman insensitive to magnetic fluctuations in the applied magnetic field of 3.2 G [1]. Initially $\sim 10^9$ atoms are trapped and cooled in a magneto-optical trap, and further cooled by Sisyphus and Raman sideband techniques. The technique of rapid adiabatic passage with constant RF radiation and a ramped magnetic field is then used to transfer the atoms from state $|5^2 S_{1/2}, F = 1; m_f = 1\rangle$ ending with 80% of the atoms at $|1\rangle$, and the rest in state $|5^2 S_{1/2}, F = 1; m_f = 0\rangle$. The atoms are loaded into an optical dipole trap created by two horizontal crossing beams at an angle of $28°$, creating an oval trap with an aspect ratio of 1:3.9. The 50 μm waist laser beams originate from a single frequency Ytterbium fiber laser at 1064 nm. Their polarization is parallel to the magnetic field, and frequency differ by 120 MHz to eliminate standing waves. The external control is composed of RF radiation at 2.15 MHz and microwave radiation at ~ 6.8 GHz, both locked to an atomic standard. The state populations is measured by recording the fluorescence of a detection beam resonant with a transition to an excited $5^2 P_{3/2}$ state [2]. An effect which should be taken into account is the levels shift induced by the MW field [3]. We have carefully measured this shift, which in the maximum MW power reaches to ~ 50 Hz, and it is taken into account when setting the frequency of the external fields. We carry out evaporative cooling for 2.5 s to a

Y. Sagi, *Collisional Narrowing and Dynamical Decoupling in a Dense Ensemble of Cold Atoms*, Springer Theses, DOI: 10.1007/978-3-642-29605-5_2, © Springer-Verlag Berlin Heidelberg 2012

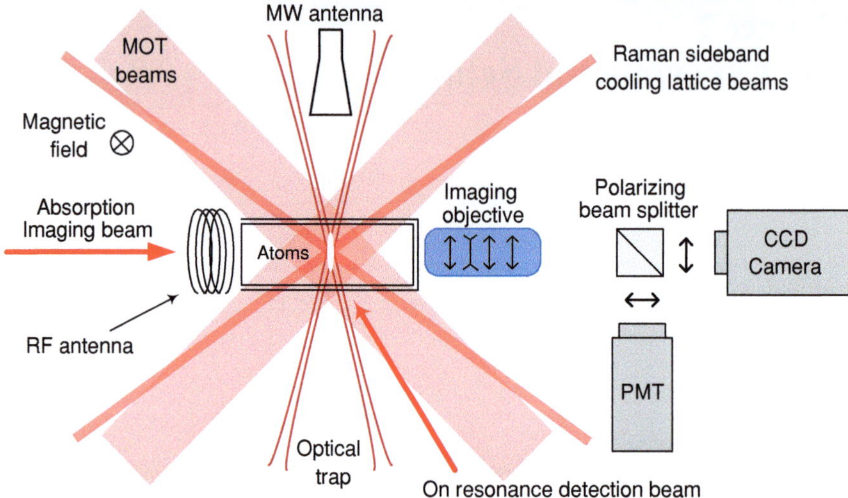

Fig. 2.1 We laser cool ^{87}Rb atoms and trap them in a crossed red-detuned laser beams configuration. We employ state sensitive detection using a detection beam and a photomultiplier tube (PMT), and measure the density and temperature using absorption imaging on a CCD camera

laser power of $0.16\,\mathrm{W}$ and back to the final laser value which determines the thermodynamic conditions in the experiment. The typical spontaneous scattering rate is less than $1\,\mathrm{s}^{-1}$ and the trap lifetime is longer than $5\,\mathrm{s}$. The temperature is measured by an absorption imaging after a time of flight.

2.2 General Considerations

The design of the new setup is based on several considerations. The two main goals are that the new setup will work with high data acquisition rate, and that it will enable good optical access to incorporate the optical lattice beams, detection and trapping lasers. In conventional BEC setups based on two chambers and magnetic trapping the evaporation sequence takes around a minute, and the typical data acquisition rate is therefore a data point every $\sim 100\,\mathrm{s}$. The long evaporation times are a consequence of the low collision rates and low densities in the magnetic trap. In order to increase this rate we have decided to use an optical dipole trap instead. The natural densities and collision rates in an optical trap are much higher, and this allows for much faster evaporation, typically on the scale of a few seconds [4, 5]. The short evaporation times ease the vacuum requirements, and enables us to work in a single chamber. Since we do not use magnetic trapping, it is not necessary to work with very short distances which are usually required to obtain large gradients of the magnetic field. Thus, the size of the vacuum chamber can be larger, which provides a better optical access. Another advantage of the larger size is that one can use large diameter MOT beams, which increase the number of atoms collected into the MOT for a given ^{87}Rb background vapor pressure. We have designed the MOT coils in such a way that we gain optical access to the atoms at $45°$, which then can be used to construct a 3D cubic optical lattice for all sides of the chamber.

To further improve the initial conditions for the evaporative cooling, we have implemented a third optical cooling scheme, Raman sideband cooling [6, 7], in addition to the conventional MOT and polarization gradient cooling schemes (PGC). The details of the implementation and the scheme will be given later, but in short, it works for 10 ms, during which the phase space density of atoms is increased by two orders of magnitude and the temperature is lowered to $\sim 1.5\,\mu K$.

The MOT beams have a diameter of 30 mm and are retro-reflected by a cat-eye setup. The total power of the MOT beams is ~ 650 mW (adding the power of the retro-reflected beams), and it is derived from an external cavity diode laser amplified by a tapered amplifier. The repump power is ~ 20 mW. All lasers are prepared on a separate table and delivered to the setup by optical fibers. Around the chamber there are three pairs of compensating coils through which a current is tuned to cancel residual magnetic fields at the location of the atoms. Very good cancelation is attained though the use of MW spectroscopy of the Zeeman levels. The residual magnetic noise in the lab was measured to be 5 kHz. Rubidium emerges from dispensers operated in a steady state current of 2.4 A, chosen as a tradeoff between longer vacuum lifetime and smaller number of atoms in the MOT. For a given ^{87}Rb background pressure, we use ultra-violet LEDs which desorb atoms from the chamber walls and increase the number of atoms in the MOT and the loading rate [8, 9].
A typical experimental sequence is composed of the following stages:

- Three seconds of Magneto-optical trap loading. Typically 10^9 atoms are collected.
- 30 ms of compressed MOT phase. Magnetic field is not changed, but the repump intensity is lowered and the cooling light frequency is shifted to the red.
- Two consecutive PGC phases, separated by 5 ms, each is 2 ms long.
- Raman sideband cooling phase, 12 ms long.
- During all this time the dipole trap is on, and at typically $t = 3.023$ s all cooling lasers are shut off.

2.3 Magnetic Coils

The design of the new set of coils had to reach the necessary anti Helmholtz fields required for the MOT operation (typical gradients are 10 G/cm), be able to produce in a Helmholtz configuration a magnetic field of 1007 G required for a Feshbach resonance in ^{87}Rb, and finally preserve the optical access at $45°$ relative to the axis connecting the centers of the two coils for future use (i.e. in constructing optical lattices). The new design is based on two coils with 49 windings each of copper tubes with an outer diameter of 1/8". To comply with the third requirement we used

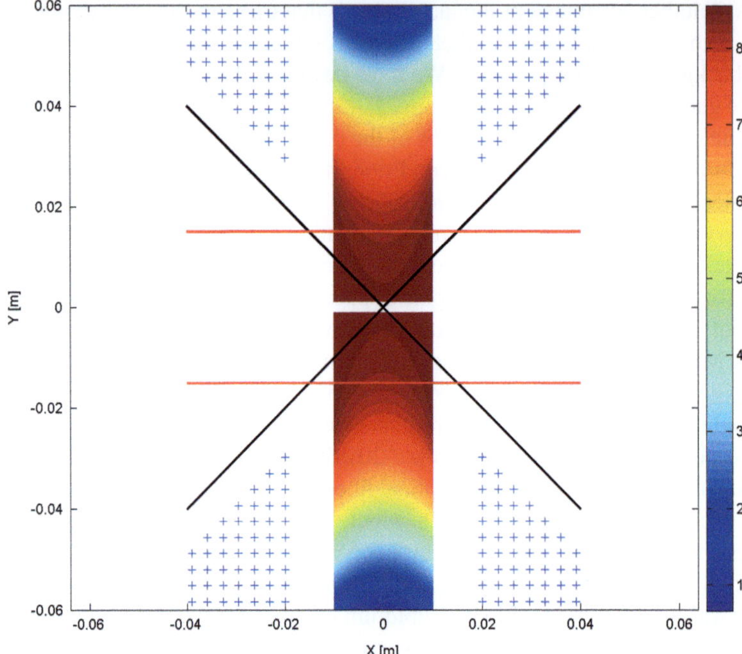

Fig. 2.2 Magnetic field magnitude [G] in a Helmholtz configuration of the coils with 1 A current. The *blue* markers are the cross sections of the coils windings. The *black lines* in the figure are the optical axes at 45° to the coils axis. The *red lines* depict the diameter of the MOT beams

Table 2.1 Physical characteristics of the new coils

	Calculated	Measured
Current to produce magnetic field of 1007 G	120 A	126 A
Resistivity of a single coil	0.09 Ω	0.076 Ω
Optimal current for MOT	10 A	8 A
Exiting water temperature at $I = 100$ A	60°	80–60°

a conical hollow center for the coils. The design and the Helmholtz fields created by a current of 1 A are shown in Fig. 2.2. From this calculation we see that a current of 120 A will be needed in order to achieve the 1007 G Feshbach resonance. The use of tubes enabled us to use cooling water, which is available in the lab, to cool the coils efficiently at such high currents. In Table 2.1 we compare the designed and achieved values of several physical characteristics of the new coils.

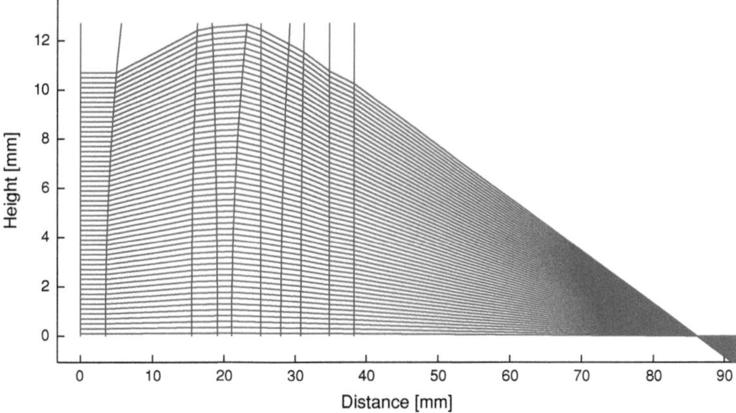

Fig. 2.3 Raytracing of the designed objective. The *red lines* are the surfaces of the lenses. The final set of *red lines* is the chamber wall

2.4 Imaging Setup

The imaging setup is a conventional on resonance absorption imaging setup [10] with two unique features: First, the imaging setup objective lens was designed and built according to our calculations, and it gives a resolution of $\sim$3.3 µm with a relatively large working distance of 50 mm. Second, a new technique was developed in order to take a reference picture in less than 100 µs after the absorption picture with the atoms. With such a short waiting time, there is a substantial reduction of the noise in the analyzed picture. The technique relies on an intense pumping of the atoms into a far detuned state before taking the reference picture. The pumping process is very short, and therefore enable us to take pictures 40 µs long and separated by only 50 µs. This effectively freezes the movement of the fringes which usually appear in the absorption beam and produces a background which is almost shot noise limited.

2.4.1 The Imaging System Objective

The heart of the imaging system is the objective lens. Due to the sizes of the chamber, we wanted a diffraction limited lens, with a working distance of at least $\sim$50 mm, a diameter of an inch and preferably a lens which is made of standard optical elements (and thus cheaper). We have written a raytracing script in Matlab (the code is given in Appendix B), and used as a starting point the design given in Ref. [11]. We restricted ourselves to use only standard lenses from Thorlabs, and used their published physical properties in the raytracer. Our final design is given in Fig. 2.3. The lenses used in this design are LC1582 LB1676 LA1608 LE1202, and the distances and properties are given in the Appendix. The designed objective has a NA of 0.2 and spherical

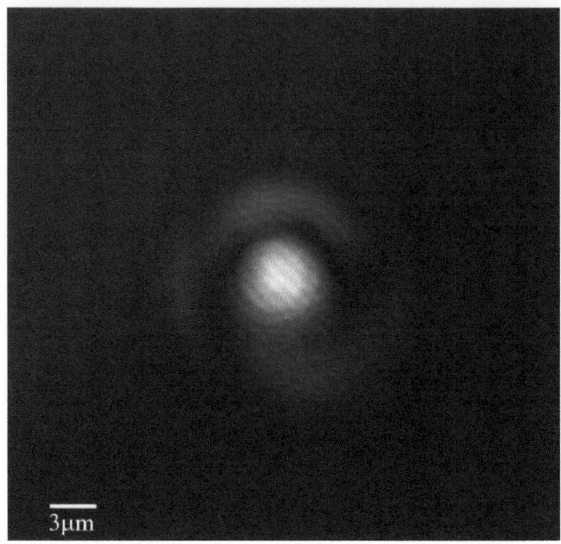

Fig. 2.4 Airy pattern measured for the home built objective

geometric abberations much less than the diffraction limit, which is $\sim$2.5 µm at a wavelength of 780 nm. The designed effective focal length of the objective is 50.8 mm.

We have built the objective using mounting tubes and rings we had bought from Thorlabs. We have tested the resolution of the objective with the following setup; 780 nm collimated light emerged from a single mode fiber (waist $\sim$1.1 mm). Afterwards it was enlarged by a telescope of 7.5–50 cm lenses (magnification of $\times$6.66), and injected into the objective. We made sure that the objective is indeed parallel and centered to the beam. After that we have placed a 3.7 mm long window to simulate the vacuum chamber wall. Afterwards, we used a commercial x60 objective held on a xyz translation stage and imaged the spot onto a calibrated camera. The measured picture is given in Fig. 2.4. The measured resolution (half the Airy disk) is $\sim$3.3 µm.

2.4.2 Absorption Imaging with Very Short Delay Times Between Pictures

Absorption Imaging is usually the main diagnostic in BEC experiments. Two images are taken, in the first one the light interacts strongly with the atoms (i.e. absorption) whereas in the second not. The background is then subtracted from the two images, and the absorption signal can be calculated. Each picture of the absorption beam contains many fringes arising from interference from parallel planes in the way of the absorption beam (i.e. windows, vacuum chamber, lenses, and so forth). These fringes tend to change their position at acoustic timescales, which results in a structured noise left in the picture which is not related to information from the atoms. In order to eliminate this noise it is necessary to reduce the time delay between the pictures. Nevertheless, going down to short delay poses an inherent problem: the atoms do not move fast enough to be absent from the second picture. There are numerous possible solutions for this problem, and we are going to describe here one of them which we have implemented successfully in the setup.

The atoms can initially be either in the $F = 1$ or $F = 2$ state. The absorption beam is resonant with the $F = 2 \rightarrow F' = 3$ transition, and thus it is necessary to add to the first pulse also a repump beam resonant with the transition $F = 1 \rightarrow F' = 2$. After the first absorption pulse is over (usually its duration is 50 μs or less), we switch on a depumping beam resonant with the transition $F = 2 \rightarrow F' = 2$. This beam is on for only a short time (typically less than 50 μs), and it efficiently drives the atoms into the $F = 1$ state. Then, only 100 μs after the first pulse, a second pulse is switched on, but this time it only includes the absorption beam and not the repump. Since ideally there are no atoms in $F = 2$ there will be no atomic absorption in this pulse, and it will serve only as the background picture. Since the repump in the first pulse is not collinear with the absorption beam, only the absorption beams enters the camera directly. The intensities of the two pictures is on the average the same. The differences that still exists arise due to the fact that the camera technically can't close the shutter in the second image quickly and therefore collects more background noise.

To demonstrate the technique we cool and trap ^{87}Rb in a dipole trap. The trapping laser is shut off exactly before the absorption pulse time in order to avoid substantial *Stark* shifts. The imaging consists of two 50 μs pulses as described before, separated by a time delay of 50 μs in which a depumper beam is switched on. The imaging optics is constructed by the objective described earlier and a second $f = 150$ mm lens that image the picture into a *PCO Pixelfly qe* double shutter camera with resolution of 1024×1392 pixels. With this optical setup, each pixel correspond to 2.24 μm in the plane of the atoms.

The raw images are shown in Fig. 2.5. There are many fringes and inhomogeneities in the two pictures, but the atomic absorption signal is still apparent. The intensity in the two pictures is different by only 2%. For weak absorption beams ($I \ll I_{sat}$), the pictures are analyzed by first subtracting the relevant dark pictures (taken before the experiment without lasers), dividing the two pictures with the appropriate intensity correction (so they have the same average intensity), and finally a logarithm is taken. The result is the optical density of the atoms, an example of which is shown in Fig. 2.6. A zoom of the atoms in the trap is shown in Fig. 2.7. The standard deviation of the background noise is $\sim$0.048 (in the optical density), which corresponds to $\sim$8 atoms per pixel. The shot noise of the laser beam can be calculated in the following way. In a pulse duration of Δt we calculate the number of photons n that incident on a single pixel. The shot noise in the optical density is than given by $\log{(n + \sqrt{n})}/(n - \sqrt{n}) \approx \log{1 + 2/\sqrt{n}}$. For a 50 μs pulse of 100 μW laser, and taking into account the quantum efficiency of the camera (25%), this results in a noise of $\sim$0.035 in the optical density. All this assumes a uniform distribution of the

Fig. 2.5 Raw images taken in the absorption scheme described

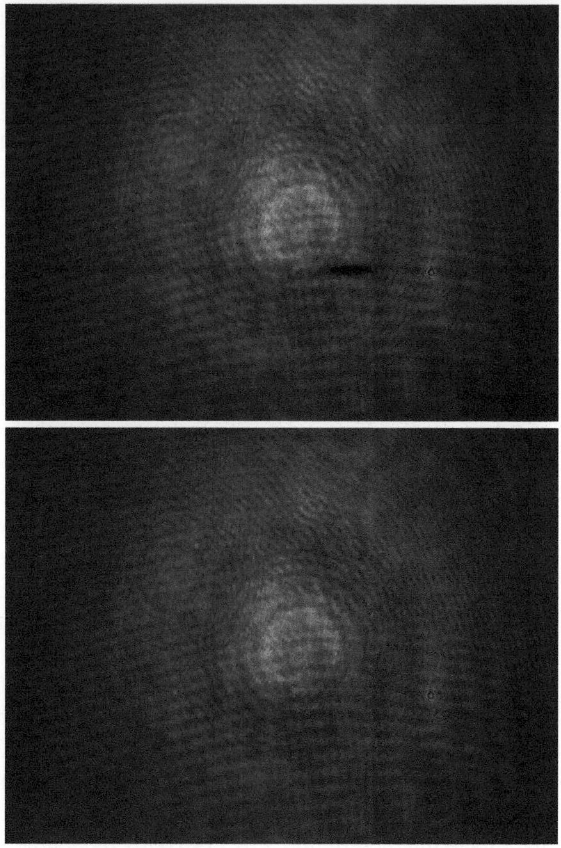

laser on all pixels. We see that the noise in the absorption picture is 40% higher than the shot noise limit.

When the optical density is substantially higher than unity, there is an advantage to work with absorption beam with higher intensity [12]. In this case, a more careful analysis of the absorption picture is required. Let us denote by I_{in} the incident absorption light, than the number of photons scattered by an atom in one second in given by $\Gamma_{sc} = \gamma/2 \cdot (1 + I_{sat}/I)^{-1}$, where I_{sat} is the saturation intensity, γ is the natural linewidth of the transition, and we assume an on resonance light. Notice that if the optical density is high, and if I is comparable to I_{sat}, there will be an exponential decrease in the transmitted intensity, which in turn modifies the scattering rate. This effect is the deviation from the usual analysis described before. We can write a differential equation describing the propagation of the absorption beam intensity through the atoms:

$$\frac{\partial I}{\partial z} = -\rho \frac{\hbar\omega\gamma}{2} \frac{I}{I + I_{sat}},$$

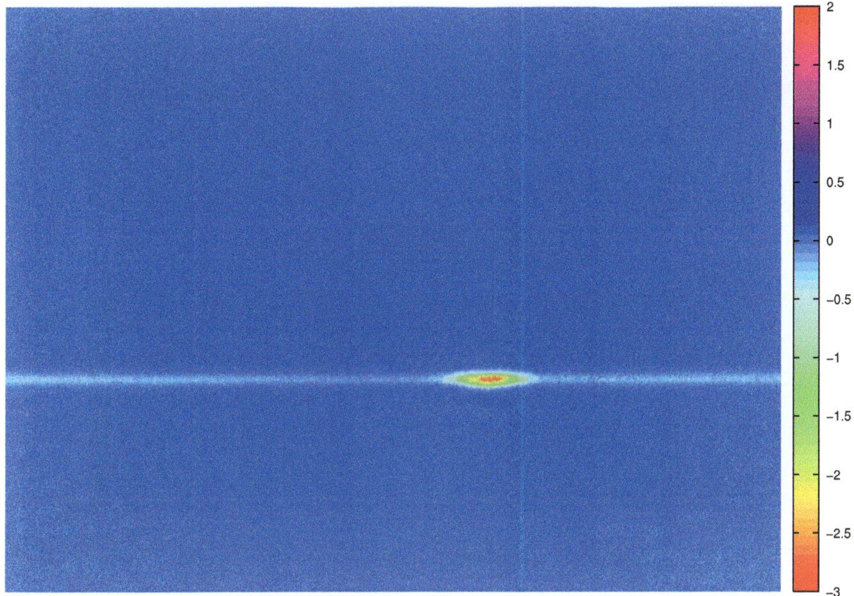

Fig. 2.6 Absorption image of atoms in a 80 μm dipole trap. The colors represent the optical density. The crossing high optical density oval shaped cloud is the crossing region of the two crossed trapping beams. One can see that many atoms are also captured in the "wings" of these two beams

where ω is the transition frequency, ρ is the density of the atoms and z is the propagation direction. We shall multiply the equation by $(1 + I_{sat}/I)$ and integrate along z to arrive at the result:

$$\frac{I_{out} - I_{in}}{I_{sat}} + \ln(I_{out}/I_{in}) = -\rho_s \frac{\hbar\omega\gamma}{2I_{sat}},$$

where we define the surface density by $\rho_s(x, y) = \int \rho(x, y, z)dz$, and use the notations I_{in} and I_{out} for the incoming and outgoing beams, respectively. The calibration of the camera enables us to translate the recorded picture into intensity, and because the intensity difference between the two pictures is very small (normally no more than 2%) we can assume that the absorption pictures gives I_{out} and the reference picture gives I_{in} (in both picture we subtract a "dark" picture without the beams to eliminate any systematic background noise). Once we have I_{out} and I_{in} we can plug it into the equation and obtain $\rho_s(x, y)$. Since the system has a cylindrical symmetry, from $\rho_s(x, y)$ we can infer $\rho(x, y, z)$. This is especially easy if we assume a certain shape of the cloud, i.e. a Gaussian.

From the measurement of the density distribution in different times we can measure all other parameters; temperature can be measured by the time of flight technique, oscillation frequency can be measured by observing the shape of the atomic cloud after some disturbance is made in the trap (more details on this later) and finally the

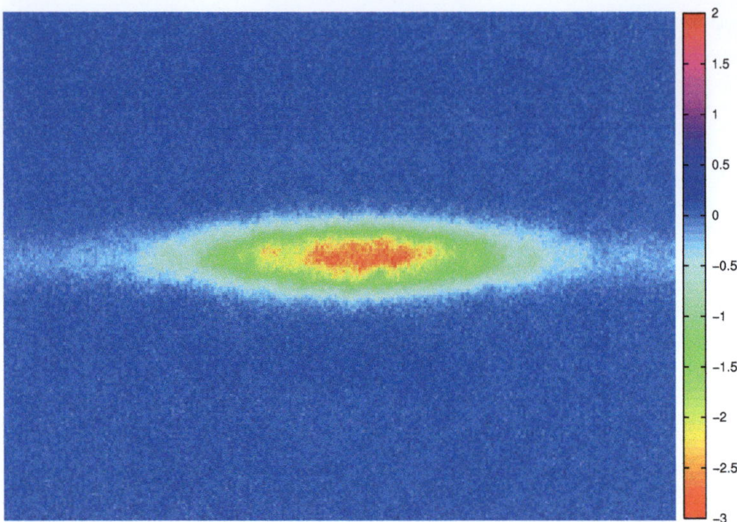

Fig. 2.7 Absorption image of atoms in a 80 μm dipole trap, zoomed. The colors represent the optical density

total number of atoms can be found by integration over the density distribution. From measurements of T, ω_{osc}, N and ρ we can also calculate the phase space density and collision rate.

2.5 State Manipulation and Detection

In this thesis the two relevant internal states are $|1\rangle = |F = 1; m_f = -1\rangle$ and $|2\rangle = |F = 2; m_f = 1\rangle$ in the $5^2S_{1/2}$ manifold (see Fig. 2.9). The advantage of working with these two levels is that, to first order, they are Zeeman insensitive in an applied magnetic field of 3.2 G [1]. Since $\Delta m = \pm 2$ between these two states, the external control $\Omega(t)$ is done by a two-photon transition, employing RF radiation at 2.15 MHz and microwave radiation at 6.832527928 GHz (see Fig. 2.9). An effect which should be taken into account is the microwave dressing effect, namely the shift of the energy levels due to the applied MW field [3]. We have carefully measured this shift as a function of the MW power and found it can typically reach a maximum of about ∼50 Hz. This shift has to be taken into account when setting the frequency of the external fields.

We measure the Rabi frequency, Ω, of the external field in the following way; We detect the population at state $|2\rangle$ while scanning the duration of the applied external field pulse. An example of such a measurement is given in Fig. 2.8. We fit the oscillating data points with the function $y = a\cos[\Omega t] + b$ and extract the Rabi frequency. In the given example the fit yields $\Omega/2\pi = 628.5 \pm 0.8$ Hz, where the error margins are given with a confidence level of 95%. The accuracy of this measurement

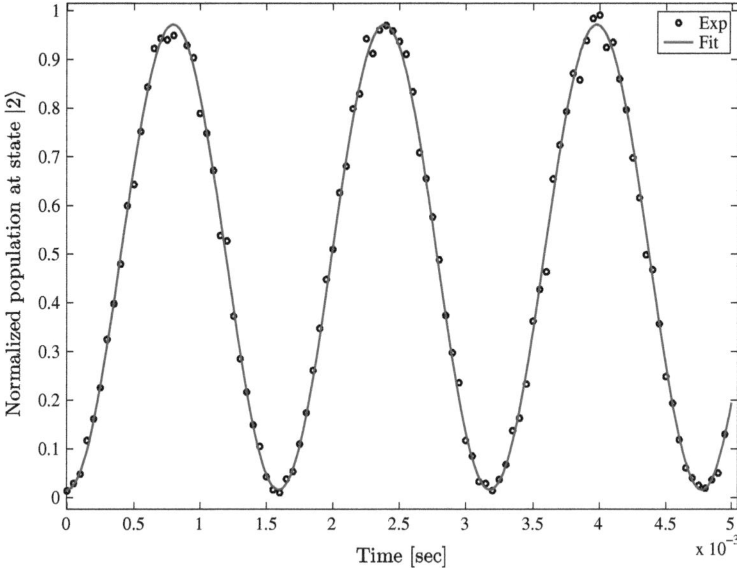

Fig. 2.8 Calibration measurement of the Rabi frequency of the control field. The x-axis is the duration of the pulse, and the y-axis is the normalized population at state $|2\rangle$. We fit the data to the function $y = a\cos[\Omega t] + b$ to find the Rabi frequency Ω

Fig. 2.9 Diagram of the hyperfine ground states of ^{87}Rb, with the two relevant magnetically insensitive states. A two-photon transition is employed, with a large enough detuning from the intermediate level

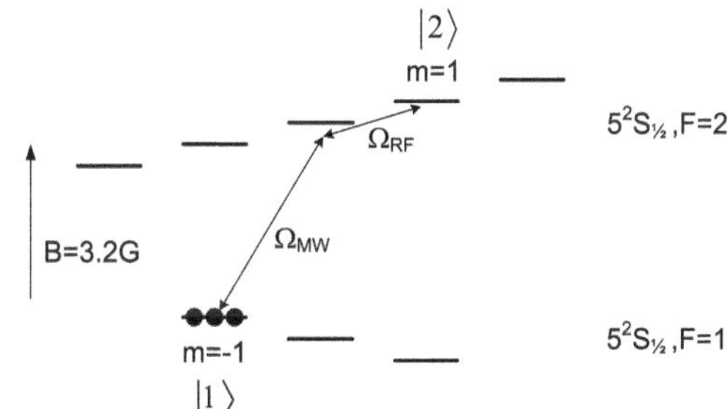

is 0.13%. Also note that the time resolution we have in setting the external pulse duration is 25 ns. Spatial variation of the external control field over the ensemble are very small since the wavelength of the microwave radiation is $\sim$4.4 cm compared to the typical size of the atomic cloud which is $\sim$100 μm. The single π-pulse fidelity was measured to be $F = 0.995$ in an experiment in which we induce more than 100 sequential pulses and measure the oscillations decay.

The state detection scheme is similar the one described in Ref. [2]. We use a detection laser beam which is resonant with the transition $|5^2S_{1/2}, F = 2\rangle \rightarrow |5^2P_{3/2}, F' = 3\rangle$. We first give a 1 ms long detection pulse which probes the population in the state $|2\rangle$. We measure the fluorescence signal from the atoms using

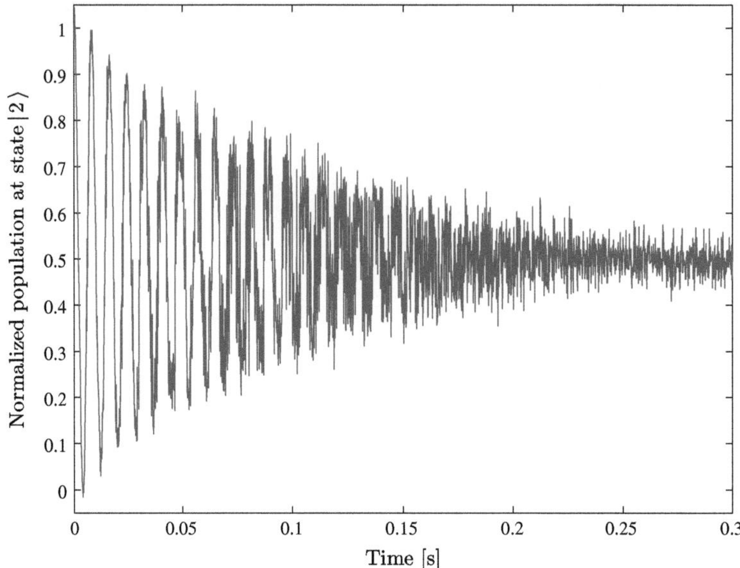

Fig. 2.10   A typical result of a Ramsey experiment with $\Gamma = 130\,\mathrm{s}^{-1}$. The duration of a $\pi/2$ pulse is $250\,\mu\mathrm{s}$

a photomultiplier tube. This detection pulse also gives the atoms at $|2\rangle$ momentum kicks which drive them out of resonance due to the doppler shift. We then give a short pulse of a repump laser which is resonant with the transition $\left|5^2S_{1/2}, F = 1\right\rangle \rightarrow \left|5^2P_{3/2}, F' = 2\right\rangle$. The purpose of this pulse is to transfer the atoms at state $|1\rangle$ to $\left|5^2S_{1/2}, F = 2\right\rangle$. Finally, we give another 1 ms long detection pulse which detects the atoms that were is $|1\rangle$. We normalize the signal of atoms at state $|2\rangle$ to the total signal, and we obtain the probability to find an atom at $|2\rangle$. Note that this three pulse sequence is carried in the same experimental run.

As will be explained in the next chapter, we use a time-domain Ramsey experiment to measure the coherence; a short $\pi/2$ pulse produced by the external control field prepares the atoms in a superposition $|\psi\rangle = \frac{1}{\sqrt{2}}(|1\rangle + |2\rangle)$ followed by a waiting time t, then a second $\pi/2$ pulse and finally a measurement of the population at $|2\rangle$. A typical result can be seen in Fig. 2.10. Such a measurement consist of 2000 points taken with different waiting times between the two $\pi/2$ pulses. The fast oscillations are a manifestation of the detuning between the transition frequency and the control field. To facilitate the extraction of the decoherence envelope function, we set this detuning to be much larger than the decay rate which ensures a separation of timescales. The envelope function is extracted by taking the standard deviation of n points, where n is the average number of points in a single fast oscillation period.

2.6 The Trap Setup

The heart of the new setup is the optical dipole trap. The use of a far detuned dipole trap has many advantages; The depth of the trap depends only weakly on the Zeeman sublevel of the atoms. The initial conditions of the atoms are much more favorable than those in a magnetic trap, probably due to a local increase in the phase space because of the dimple effect [13], and also suppression of re-scattering. Another advantage is the high initial collision rate, which in turn enables shorter evaporation times. Nevertheless, the main disadvantage is the lack of a similar effect to the so called 'RF knife' in a magnetic trap [10], namely the removal of the hottest atoms in a controlled way without changing the oscillation frequency of the trap. In a gaussian beam trap, the radial oscillation frequency can be approximated by $\omega_{osc}^2 = 4U_0/m\sigma^2$ where U_0 is the trap depth, m is the mass and σ is the beam waist (e^{-2} radius). The common and easiest way to force evaporation is to decrease the power of the laser [4]. Since this procedure also lowers the trap depth U_0 it decreases the oscillation frequency. Thus, the efficiency of the evaporation deteriorate as evaporation progress and it is not possible to carry runaway evaporation (evaporation process in which the collision rate increases). One way which was proposed to circumvent this problem is to dynamically change the waist of the beam during the evaporation [5].

In order to change the waist of the laser we have designed and built an optically compensated zoom system. Also, to keep the aspect ratio of the trap constant we have decided to work with a two beam trap in a crossed configuration. The aspect ratio is directly controlled by the crossing angle between the two beams. The trapping laser setup is depicted in Fig. 2.11. The trap laser is an Ytterbium fiber laser manufactured by IPG, model YLR-50-1064-LP-SF. This is a 50 W single frequency fiber laser (linewidth < 50 kHz) with a wavelength of 1064.347 nm and a waist of 1.5 mm. We tested the new laser with a high speed detector (8 GHz BW) and verified that there are no more than a single mode, at least at the measured bandwidth. The setup produces a 28° crossed beam trap with a waist range of 50–250 μm and an aspect ratio of 1 : 4. The total power reaching the atoms is ∼30–32 W. The laser emerges with a linear polarization through an optical isolator. It is than passed through a telescope which is focused to a waist of 800 μm on an AOM (Model 35060-30-3-1.06-I-HGM-W with driver model 39060-30DMA05-A, manufactured by NEOS Tech). After the telescope there is a polarization beam splitter (PBS1) and a mirror (M1) which create the two beams that are going to enter the AOM. The ratio between the two beams can be controlled using a $\lambda/2$ placed just after the laser head. The two beams are aligned such that they heat different spots on the AOM. The right (left) beam is resonant with the moving acoustic scattering lattice in the AOM and diffracts into the order $+1$ (-1). The remaining zero orders of both beams are removed by a beam dump. The two diffracted beams are then mode matched and recombined on a beam splitter (PBS2). Since the AOM is sensitive to the incident polarization we rotate the polarization of the left beam before and after the AOM. The AOM operates at 60 MHz, therefore the frequency difference between the two diffracted beams is 120 MHz which effectively prevents effects of standing waves. Mirrors M3 and M4 are used to align the beam

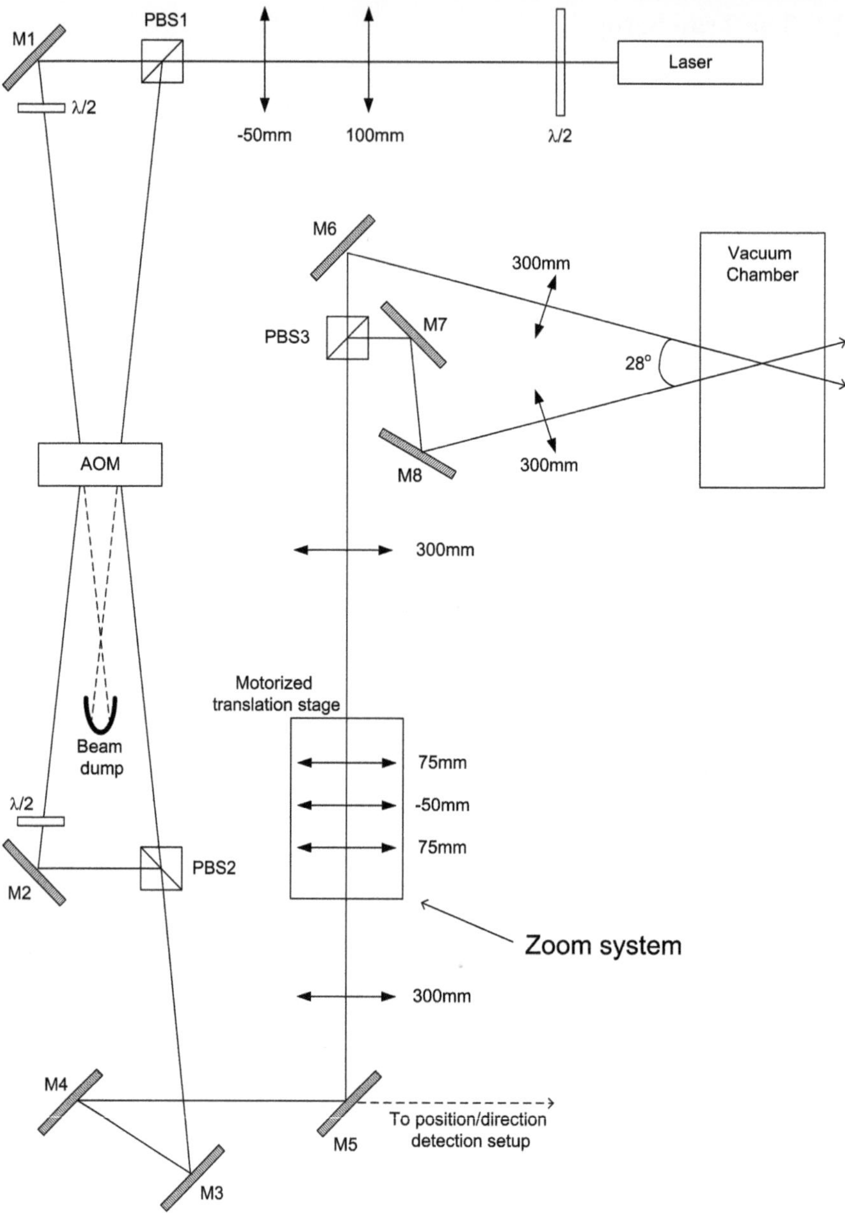

Fig. 2.11 A sketch of the experimental setup of the trapping laser

to a specific mode which is measured by two quadrant detectors placed after the leakage of mirror M5. In case of problems before Mirror M5, this enables us to correct the problems and then realign very accurately the beam such that it does not require further realignment after M5. The lenses after M5 create the zoom system (it will be described afterwards in more detail). The two beams are finally separated by polarizing beam splitter (PBS3) and sent into the chamber. The final polarization state of the two beams can be further manipulated by waveplates just after mirrors M6 and M8. The holders of mirrors M6 and M8 include piezoelectric elements that enable real time fine tuning of the crossed position. After the trap beams leave the chamber we image them on a position sensitive quadrant detector, and lock their position using the piezoelectric mirror mounts. The position locking corrects small displacements of the focus position during the dynamic compression of the zoom. Also, we continuously monitor the beams power (we do this also at the leakage after M5), and lock the power to the desired value. This value changes during the evaporation, and so the power locking system keeps the beams power locked to this value dynamically.

In an ideal zoom system the beam waist changes without affecting its position. However, to a certain extent a movement of the waist is acceptable if it is kept under some limitations. At the crossing point of the two trapping beams the potential depth is twice the depth of each single beam. If the waist is more than one Rayleigh range away from the crossing point, another local minima of the potential appear at the waist of each beam. Therefore, it seems reasonable to tolerate a movement of the waist position up to one Rayleigh range of the beam.

It is well known that an ideal zoom system requires an independent movement of at least two optical elements (mechanically compensated zoom systems). There is a different approach which require only a single moving platform (so there is only a single independent moving variable), but allow the concurrent movement of several optical elements. This approach gives an approximation of a zoom system, and is called optically compensated zoom system [14]. In the lowest order of implementation we allow the concurrent movement of two lenses, where in between there is a static third lens (the moving lenses does not change their relative distance).

We have designed optically compensated zoom system which achieve a waist range of $50-250\,\mu$m. Design parameters were optimized using Gaussian ABCD matrices simulations. Our design is based on two positive moving lenses with focal length of 75 mm positioned 10 cm apart from each other, and in between one fixed negative lens with a focal length of -50 mm. The result of the simulation of our design is given in Fig. 2.12. We have built this system and verified that indeed we get the designed waist range.

2.7 Raman Sideband Cooling

Raman sideband cooling is an optical cooling scheme first proposed and demonstrated by Steven Chu's group in 1998 [6, 7]. In this scheme, one uses a magnetic field to shift the relative energy of two vibrational ladders of two Zeeman sub levels such that the states $|m; n\rangle$ and $|m + 1; n - 1\rangle$ are degenerate (m is the Zeeman index and n is the vibrational state index). Once the levels are degenerate, Raman

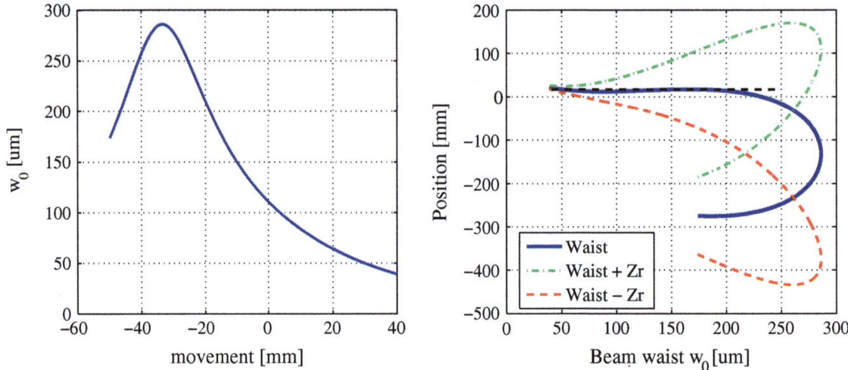

Fig. 2.12 Simulation of a three component optically compensated zoom system. Simulation are done using ABCD matrices. *Left* graph is the waist size as a function of translation stage position. We can verify that we have a factor of over six magnification of waist size for 85 mm of motor travel. *Right* graph is the waist position as a function of the waist size. The position of the ±Reighley range from the waist position is also depicted. The *black dashed line* indicate that we can choose a fixed position in space in which our criterion of waist position movement is fulfilled

transitions move the atoms from $|m = -1; n + 1\rangle$ to $|m = 1; n - 1\rangle$. In this transition the atoms loose two vibrational quanta. The final stage of the scheme is optical pumping back into the $m = -1$ state. One has to make sure that the confinement is strong enough such that the atoms are in the Lamb-Dicke regime, and thus the probability for a change in the vibrational level during the optical pumping process is small. To accomplish this and also to create a well separated vibrational ladder, we built a four beam optical lattice. The lattice beams also induce the degenerate Raman transitions.

The implementation of the scheme in ^{87}Rb is depicted in Fig. 2.13. Magnetic field of $\sim$200 mB is applied in the z axis. The Raman lattice is composed of four beams in exactly the same geometry and polarizations as in Ref. [6]. The total power of the raman lattice beams is 350 mW and the frequency is detuned by +13 GHz with respect to the $D2\ F = 1 \rightarrow F' = 2$ transition. The optical pumping is done with a strong $\sigma+$ ($\sim$20 μW) and a weak π beam ($\sim$2 μW), both detuned +10 MHz with respect to the $D2\ F = 1 \rightarrow F' = 0$ transition. Also needed is a repumper which pumps back atoms transferred to $F = 2$ state. All beam waists are $\sim$1.1 mm.

The cooling sequence starts just after the second PGC phase. We ramp up adiabatically the lattice, and switch on the magnetic fields, optical pumping and repumper lasers for 10 ms. Finally, the lattice is ramped down, while the magnetic field is still on. A temperature measurement done in a time of flight technique is shown in Fig. 2.14. Temperatures of $\sim$1.5 μK are typical. We end up with few 10^8 atoms, in a typical waist of 500 μm, which gives a typical phase space density of 5×10^{-4}. We have also performed MW spectrometry experiments after the sequence, a typical result of which is given in Fig. 2.15. The optical pumping produces a typical distribution of 80% of atoms in $m = 1$, 15% in $m = 0$ and 5% in $m = -1$.

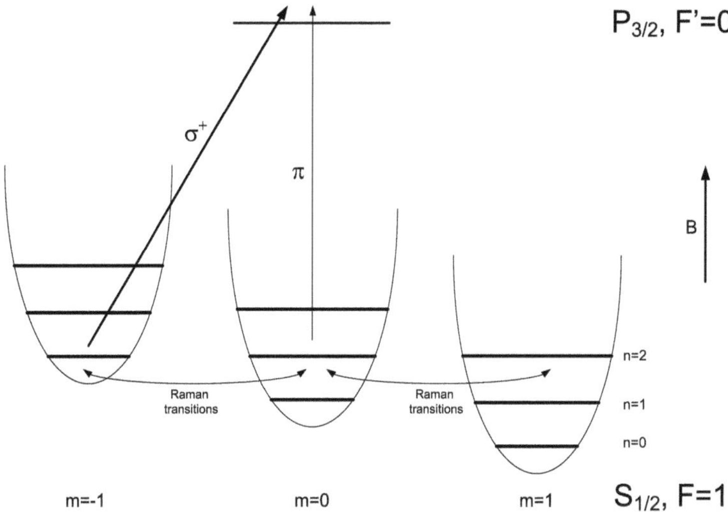

Fig. 2.13 Raman sideband cooling in ^{87}Rb atoms

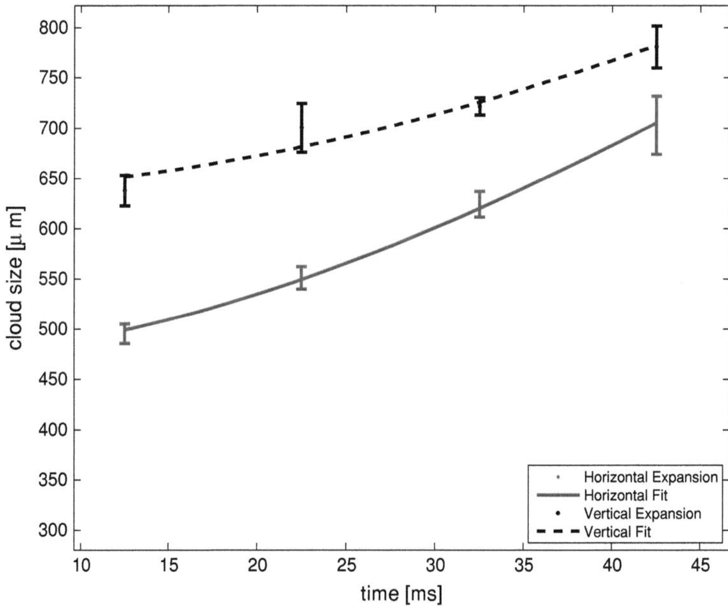

Fig. 2.14 Time of flight measurements of ^{87}Rb atomic cloud after Raman sideband cooling. Fits correspond to temperature of $1.58 \pm 0.06\,\mu$K in the horizontal direction and $1.20 \pm 0.25\,\mu$K in the vertical direction. Cloud size, σ, is found by fitting to a density distribution function: $n(r) = n_0 e^{\frac{r^2}{2\sigma^2}}$. Time is the expansion time of the cloud measured from the sudden shout down of the lattice. Error bars are standard deviations of three measurements done for each point

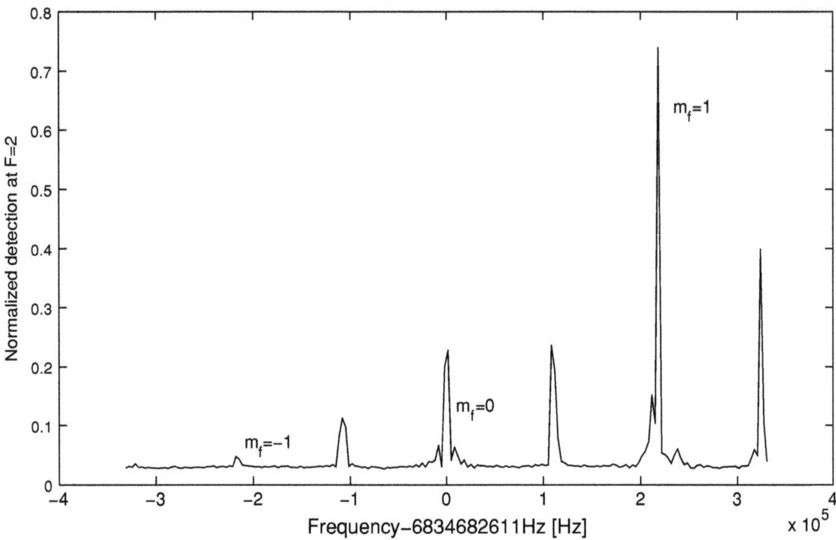

Fig. 2.15 Zeeman level distribution measured by RF magnetic transitions spectrometry. x axis is the RF photons frequency and y axis is the normalized detection signal which corresponds to the Zeeman level population. Besides the peaks of the equal m transitions there are also transition with $\Delta m \neq 0$ due to RF photons with $\sigma+$ and $\sigma-$ polarizations. In this measurement there is a magnetic field of $\sim 160\,\text{mG}$

2.8 Levitation

Raman sideband cooling enables to reach temperatures as low as $1.5\,\mu\text{K}$. The needed trap depth to load such cold atoms is very low, and thus for a given laser power larger trap waists can be used. Nevertheless, when going to larger trap waists, the gravitational energy difference across the trap can be larger than the trap depth. As an example, for a $200\,\mu\text{m}$ distance the gravitational energy for ^{87}Rb is $21\,\mu\text{K}$. A useful tool in cold atomic experiments is the use of some force to contradict gravitation and to effectively create levitation in the location of the atoms.

In our setup we have implemented levitation using magnetic forces. The main idea is to use the magnetic dipole moment of the atoms, and to create a magnetic field gradient such that the force equals mg. The magnetic dipole interaction energy is given by $H_B = m_f g_f |B|$, where we assume magnetic dipole moment adiabatic following of the magnetic field direction. For our case $m_f = 1$ (this is the resulting state after the Raman sideband cooling), and $g_f = 0.7\,\text{MHz/G}$. We use the main coils of the MOT to create the field gradient needed for the levitation. We have carried out a 3D calculation of the magnetic fields, and determined that the needed current in the coils in an anti-Helmholtz configuration is 13 A. In the experiment we have found that the value should be 12 A. The use of two coils in an anti-Helmholtz configuration facilitates the current requirement from the current driver, but the drawback is stronger

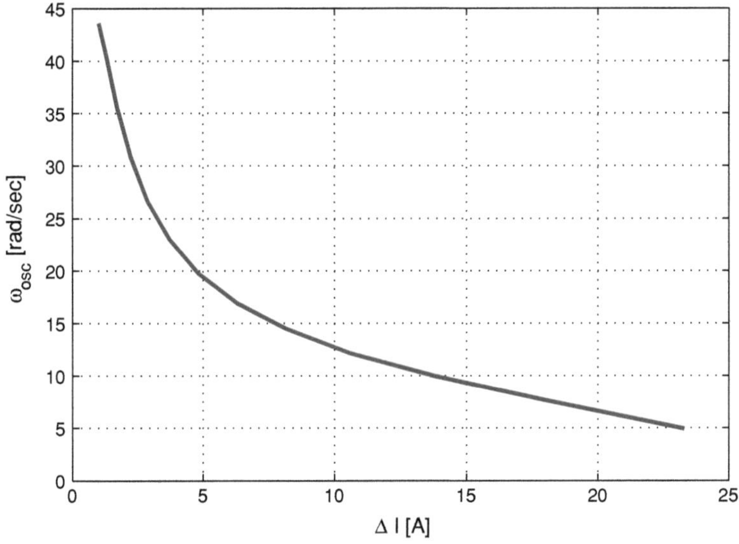

Fig. 2.16 Calculation of the transverse anti trapping oscillation frequency induced by magnetic field gradients in the transverse plain as a function of ΔI which is the difference between the current in the *upper* and *lower* coils. The sum of the two currents is kept constant at 24 A to produce the axial magnetic field gradients necessary for levitation

transverse magnetic field gradient which induces anti trapping in the transverse plain. On the other hand one can use a current of 24 A in a single coil to get better results. The calculated transverse anti trapping oscillation frequency is depicted in Fig. 2.16. We have found that the small anti trapping effect in the transverse plain can be useful to get rid of the atoms that are weakly trapped in the wings of the two trapping beams. We have found that these atoms cause excessive heating when they collide with atoms inside the crossing region. Since the transverse trapping frequencies are usually very low, these anti trapping effect helps to spill these atoms without changing effectively the potential in the crossing region.

To summarize, we have used an unbiased current in an anti Helmholtz configuration of the MOT coils to levitate the atoms. We have found and verified the exact parameters of the levitation by optimizing the number of atoms is a very shallow trap (which does not trap without levitation), and also by time of flight measurement in which we verified that there is no free fall.

2.9 Lifetime Measurements

The vacuum measured by the vacuum ion pump is $\sim 2 \times 10^{-11}$ Torr. The ambient vapor pressure at the position of the atoms is determined by the current in the Rb dispensers. This value is a tradeoff between high current which would result in more

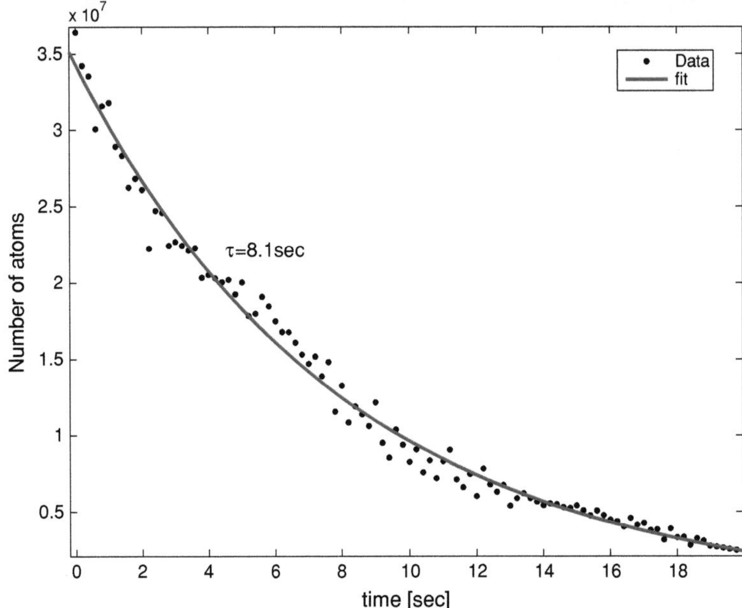

Fig. 2.17 Lifetime measurement in a magnetic trap. We fit to exponential function $N = N_0 e^{-t/\tau}$. In this measurement the atoms are prepared in $F = 1$ state

atoms gathered in the MOT and low current which will results in longer lifetimes and thus more efficient evaporation. We can measure the lifetime in different ways. One of the most sensitive ways is to measure the lifetime of atoms loaded into a magnetic trap. This is a very sensitive probe to m changing transition, because such a transition immediately transfer the atom to an untrapped state, in contrast to other heating mechanisms which require many photons to cause a loss of a single atom. Such m changing transitions occurs mainly due to stray light. We have found that with very small leakage, lifetime measured using this method are an order of magnitude shorter than as measured in other methods. The results of a typical measurement is given in Fig. 2.17, where the measured exponential decay time is 8.1 s (in this measurement the current in the getters is 2.7 A). Notice that the exponential fit describes the data well, which implies that in this measurement there is no density dependent loss. This is expected since the densities we obtain in magnetic trap are quite low. Measurements with optical dipole trap with different beam waists give similar results. Recently we have reduced the current in the getters to 2.4 A, and the lifetime increased to more than 10 s.

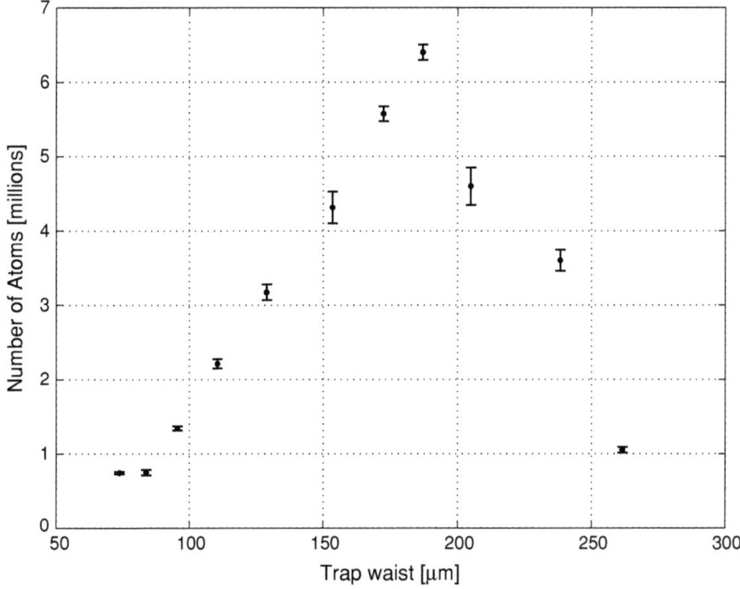

Fig. 2.18 Number of loaded atoms in the trap as a function of the trap waist. The error bars are the standard deviation of the average of seven measurements. The measurement is done 100 ms after the end of the cooling and loading phases. No levitation in this measurement

2.10 Trap Loading

The trapping laser is present throughout the optical cooling phases since we did not notice any improvement when it was shut off during this time. The number of collected atoms as a function of the trapping laser waist when the laser is at full power is given in Fig. 2.18. The graph shows that there is an optimum around $\sim$200 μm waist size. At larger waists gravitation starts to lower substantially the trap depth, and in lower trap depths the capturing volume is smaller. Actually, when trying to take these two considerations into account it seems that there is another mechanism in smaller trap waists which helps to keep the number of atoms higher than expected. We believe this mechanism to be the so called dimple effect, as was demonstrated in Ref. [13, 15]. In short, the dimple effect is a local increase of the phase space density in comparison to the phase space density of a reservoir, in a region of space where the potential is lower than the potential at the reservoir, and in thermal equilibrium. When using smaller waist traps we gain in the phase space since the potential in the trap is lower than in the Raman sideband cooled cloud. When adding levitation we gain in the number of atoms (typically not more than 20%), and the drop in the number of atoms shifts to larger waists.

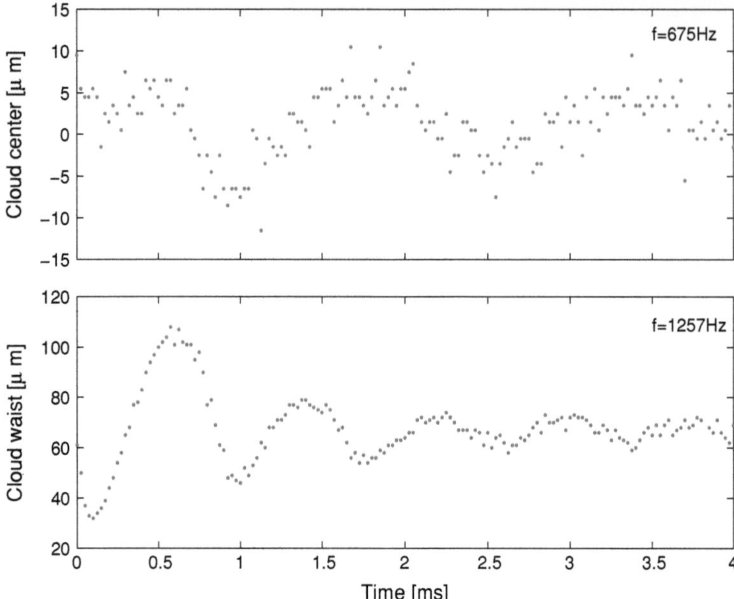

Fig. 2.19 Oscillation frequency measurement by a rapid trap switching. We switch off the trap for 1 ms and measure the cloud center (*upper* graph) and waist (*lower* graph) after 1 ms time of flight. Trap parameters are: power 15 W, $\sigma = 75\,\mu$m

2.11 Oscillation Frequency Measurements

One of the most important characterizations of the trap is the measurement of its oscillation frequency. We have employed two measurement techniques which we present here: rapid switching of the trap and parametric excitation. In both techniques, besides the oscillation frequencies there are numerous interesting physical phenomena we were able to observe such as trap non harmonicity, parametric cooling and more.

2.11.1 Oscillation Frequency Measurement by a Rapid Trap Switching

In this technique we switch off the trap for a short duration and then switch it on again and measure the atomic could geometric parameters as a function of time. The measurement is carried by absorption imaging pictures of the cloud which we fit to a gaussian and find its center and waist. The witching duration should be smaller than the oscillation period. An example of such a measurement is shown in Fig. 2.19.

The upper graph shows the oscillation of the center of mass of the cloud, and as expected its damping which is only connected to the non harmonicity of the trap is quite small. The center of mass oscillation in the trap can be written as $x = x_0 \sin(\omega_{osc} t)$, where t is the time, ω_{osc} is the oscillation frequency and x_0 is the initial displacement. After a time of flight duration of t_{tof} the measured center of mass position is given by $x_{tof} = x + v t_{tof} = x_0 \sin(\omega_{osc} t) + t_{tof} \omega_{osc} x_0 \cos(\omega_{osc} t)$ which can always be written as $x_{tof} = x_0 \sqrt{1 + (t_{TOF} \omega_{osc})^2} \sin(\omega_{osc} t + \phi)$, with some phase ϕ. We see that we may gain a "magnification" factor of $\sqrt{1 + (t_{tof} \omega_{osc})^2}$ by the use of the time of flight technique. since for a given ω_{osc} the maximum measurable time of flight duration is limited, the typical value is $t_{tof} \omega_{osc} = 10$. Still, since the displacement relative to the atomic waist is small, and since the typical atomic waist is few tenths of microns, the typical x_0 is on the order of a few microns. This results in a measurable displacement of $\sim 10\,\mu$m, which is small and thus this kind of measurement tends to be noisy.

The lower graph at Fig. 2.19, on the other hand, shows the measurement of the waist of the atomic cloud. This is the so called "breathing mode" and it is damped due to collisions. It is quite easy to induce a large relative waist change, and this translates to a measurable change of many tenths of microns. This kind of measurement tends to be easier and with better signal to noise. Nevertheless, there are two points to remember. First, as a result of the damping it is usually difficult to measure more than 10 oscillations, which limits the accuracy of the measured frequency. Second, due to non harmonicity of the trap the instantaneous frequency measured by the breathing mode shifts up as time goes on. This is because the amplitude of the motion decreases and due to non linear effect of the oscillator the frequency increases. On the average it means that this measurement technique has a small bias towards lower frequencies. Saying all that, it is still our preferred way of measuring the oscillation frequency due to the good signal to noise and accuracy.

From the damping rate of the breathing mode we could infer the collision rate directly. This is appealing since direct measurements of this value are not easy to perform [16]. The main issue which needs to be taken into account here is the effect of anharmonicitiy of the trap. The slow decay of the center of mass mode, however, shows that at our typical collision rates this effect is rather small. More details on this technique are given in Chap. 5.

2.11.2 Parametric Excitation of Atoms in a Dipole Trap

Another technique to measure the oscillation frequency is by modulating the trapping laser intensity and measuring the response of the atoms. In the experiment, we give a modulation at a certain frequency and measure the number of remaining atoms and their temperature. In a perfect harmonic oscillator we expect to see a resonance behavior at each of the oscillation frequencies of the trap. A typical measurement is given in Fig. 2.20. The temperature of the atoms show two clear peaks at the radial and

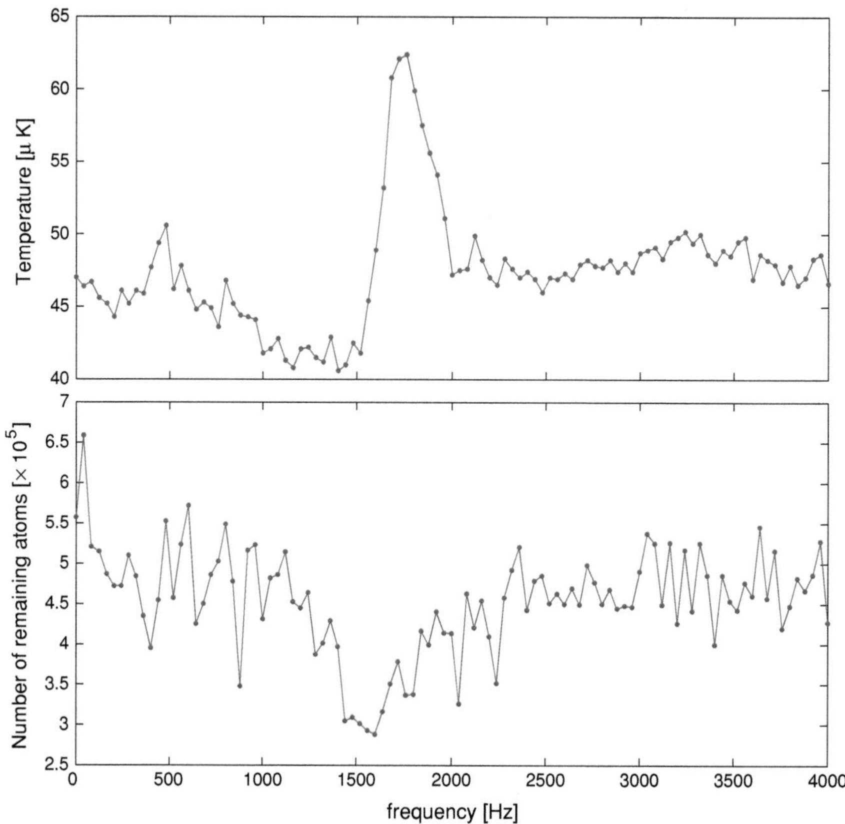

Fig. 2.20 Parametric modulation of the trap depth. *Upper* graph is the remaining atoms' temperature and *lower* graph is the number of remaining atoms. We give 100 ms of modulation which ends just before the measurement. Trap parameters: power 15 W, modulation peak to peak 3 W, waist is ∼60 μm

axial oscillation frequencies. An interesting feature is a decrease of the temperature at frequencies of 1000–1500 Hz (the temperature without any modulation is ∼47 μK). This decrease is due to the anharmonicity of the trap. The hotter atoms fill the higher energy levels of the trap, which due to the anharmonicity also have smaller oscillation frequencies. When we induce parametric excitation with smaller frequencies than the resonance frequency we on the average excite hotter atoms out of the trap and thus the remaining atoms are colder. In the lower graph in Fig. 2.20 one can see that indeed the largest number of atoms which are lost exactly in this frequency range. This effect which we call parametric cooling was already reported in Ref. [17–19].

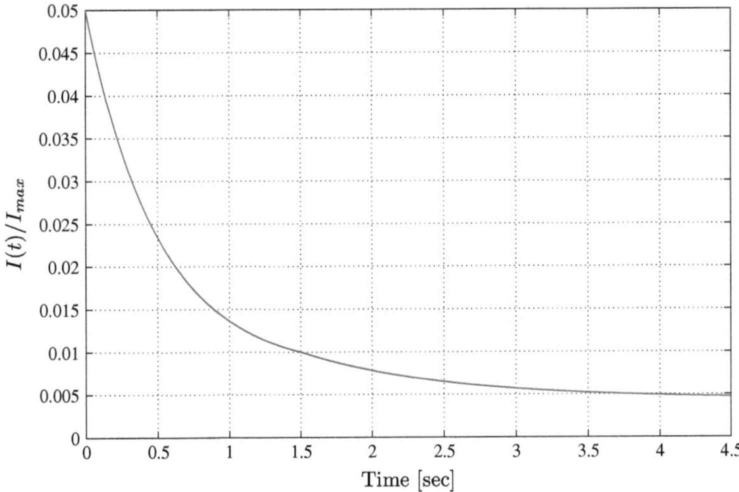

Fig. 2.21 Trapping laser power relative to its maximum value, $I_{\max} = 30\,$W, in the evaporation sequence

2.12 Evaporation to BEC

Our sequence to achieve quantum degeneracy is composed of several stages. First, we use optical cooling, as explained before, which brings the phase space of our ensemble to be better than 10^{-4}. The atoms are then loaded into the dipole trap. In order to collect a large number of atoms, we set the waist of the trapping beams to $\sim 200-250\,\mu$m and its power to maximum ($\sim$30W). We then let the atoms freely evaporate for 0.5 s. Next, we compress the trap to a waist of $\sim 50\,\mu$m during 1.5 s. This compression reduces the volume by a factor of ~ 100, and consequently the density and elastic collision rates increase dramatically. Unfortunately, also the inelastic scattering rate is increased, and we have found experimentally that the most efficient evaporation is achieved when the laser power is reduced gradually during the compression down to 5% of its maximum value. At this point there are typically 10^6 atoms at a phase space density of 10^{-2}, a temperature of 6.5 μK, a maximum density of $\rho = 3 \times 10^{13}\,\mathrm{cm}^{-3}$ and a maximum elastic collision rate of $\Gamma_{col} = 1000\,\mathrm{s}^{-1}$. This constitutes an evaporation process with an efficiency of $\sim$2.6.

Next, evaporation is continued only by reducing the power of the trapping laser. The best results to date were obtained with the following power reduction scheme: for the first 1.5 s, the power follows the function $I(t) = Ae^{-t/\tau_1} + B$ with $\tau_1 = 0.5\,$s, and A and B such that $I(0) = 0.05\,I_{\max}$ and $I(1.5) = 0.01\,I_{\max}$. For the next 3 s, the power is given by $I(t) = Ce^{-t/\tau_2} + D$ with $\tau_2 = 1\,$s and C and D chosen such that $I(1.5) = 0.01\,I_{\max}$ and $I(4.5) = 0.0047\,I_{\max}$. The total evaporation sequence is plotted in Fig. 2.21.

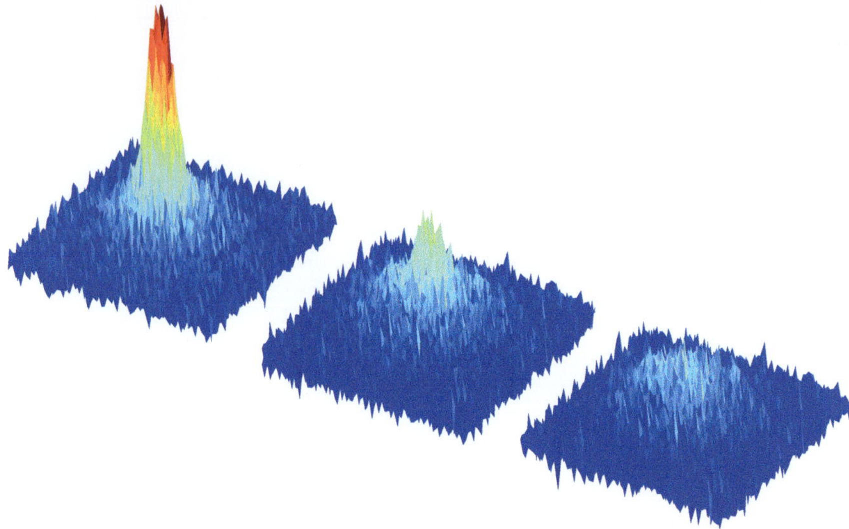

Fig. 2.22 Different stages in the condensation process, taken after a time of flight of 18 ms. The height is proportional to the optical depth in each image. In all these experiments we used the same sequence, and the different ending conditions stem from fluctuations in the initial number of atoms before the evaporation

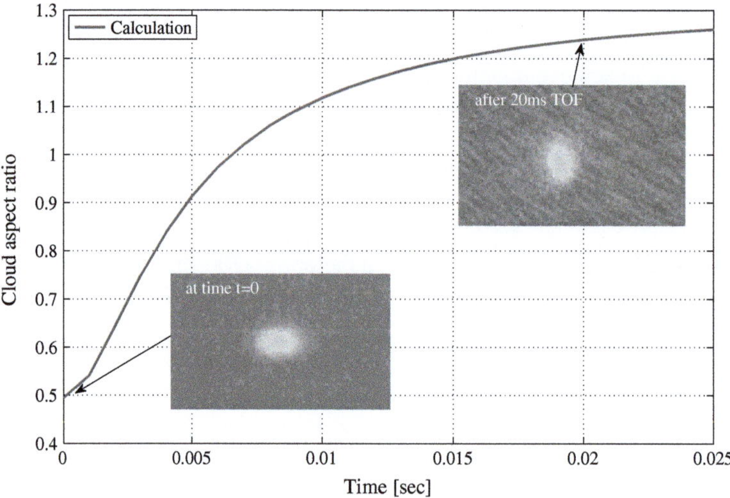

Fig. 2.23 The condensate aspect ratio calculated in the Thomas–Fermi limit using the measured oscillation frequencies. The insets show the measurements

The resulting BEC contains $\sim$5000–1000 atoms. We have measured it after a time of flight of 15–20 ms. At this point, and mostly due to the fluctuation in the initial conditions before the evaporation, the conditions after the evaporations are quite fluctuating. In future work we plan to perform the last stage of the evaporation by increasing the gradients of an applied magnetic field. Thus a higher trapping frequency can be maintained and therefore also a better efficiency. In Fig. 2.22 we depict different stages of the condensation process. In addition, we have measured the oscillation frequencies in the trap just at the condensation point and obtained $\omega_r = 2\pi \times 84\,\mathrm{Hz}$ and $\omega_z = 2\pi \times 41\,\mathrm{Hz}$. Using this values, we calculate the aspect ratio of the condensate as a function of the time of flight (in Thomas–Fermi regime) and plot it in Fig. 2.23 [10]. It is important to note that since the aspect ratio is close to unity, in the calculation of the chemical potential from experimental data the kinetic energy in both axis has to be taken into account. Doing so we find that the typical chemical potential is $\mu = 2\pi \times 300\,\mathrm{Hz}$. We extract the temperature from the thermal wings around the condensate (but still not too close) and get a typical value of $T = 50\,\mathrm{nK}$.

References

1. Harber DM, Lewandowski HJ, McGuirk JM, Cornell EA (2002) Phys Rev A 66:053616
2. Khaykovich L, Friedman N, Baluschev S, Fathi D, Davidson N (2000) Europhys Lett 50:454
3. Haroche S, Cohen-Tannoudji C, Audoin C, Schermann JP (1970) Phys Rev Lett 24:861
4. Barrett MD, Sauer JA, Chapman MS (2001) Phys Rev Lett 87:010404
5. Kinoshita T, Wenger T, Weiss DS (2005) Phys Rev A 71:011602
6. Kerman AJ, Vuletic V, Chin C, Chu S (2000) Phys Rev Lett 84:439
7. Vuletic V, Chin C, Kerman AJ, Chu S (1998) Phys Rev Lett 81:5768
8. Klempt C, van Zoest T, Henninger T, Topic O, Rasel E, Ertmer W, Arlt J (2006) Phys Rev A 73:013410
9. Aubin S, Myrskog S, Extavour MHT, LeBlanc LJ, McKay D, Stummer A, Thywissen JH (2006) Nat Phys 2:384
10. Ketterle W, Durfee D, Stamper-Kurn D (1999) Bose-Einstein condensation in atomic gases, In: Inguscio M, Stringari S, Wieman CE (eds) Proceedings of the international school of physics "Enrico Fermi", course CXL (Varenna lecture notes), p 67
11. Alt W (2002) Optik 113:142
12. Reinaudi G, Lahaye T, Wang Z, Guéry-Odelin D (2007) Opt Lett 32:3143
13. Stamper-Kurn DM, Miesner H-J, Chikkatur AP, Inouye S, Stenger J, Ketterle W (1998) Phys Rev Lett 81:2194
14. Wooters G, Silvertooth EWJ (1965) Opt Soc Am 55:347
15. Weber T, Herbig J, Mark M, Nagerl H-C, Grimm R (2003) Science 299:232
16. Monroe CR, Cornell EA, Sackett CA, Myatt CJ, Wieman CE (1993) Phys Rev Lett 70:414
17. Poli N, Brecha RJ, Roati G, Modugno G (2002) Phys Rev A 65:021401
18. Kumakura M, Shirahata Y, Takasu Y, Takahashi Y, Yabuzaki T (2003) Phys Rev A 68:021401
19. Zhang PF, Zhang HC, Xu XP, Wang YZ (2008) Chin Phys Lett 25:89

Chapter 3
Theoretical Framework

When the atoms move in the potential their energy levels slightly change. I denote the energy of the low-lying internal states $|1\rangle$ and $|2\rangle$ by $E_1(x, y, z)$, and $E_2(x, y, z)$, respectively. The depth of the optical potential is proportional to $U(x, y, z) \propto I(x, y, z)/\Delta$, where $I(x, y, z)$ is the laser intensity and Δ is its detuning from the excited states $\Delta = \omega_{laser} - \omega_i$, where ω_i is the resonance transition frequency of the state $|i\rangle$ to the excited states. Since $\omega_1 \neq \omega_2$, one gets that $\Delta_1 \neq \Delta_2$ and consequently the energy difference between the two internal states is not constant: $\Delta E(x, y, z) = E_2 - E_1 = I(x, y, z)(\frac{1}{\Delta_2} - \frac{1}{\Delta_2}) \approx I(x, y, z)\frac{\Delta_1 - \Delta_2}{\Delta^2}$, where I assume that $\Delta_1 - \Delta_2 \ll \Delta_1, \Delta_2$. In particular, the energy difference between the states has the same spacial dependence as the laser intensity and the optical potential $\Delta E(x, y, x) \propto U(x, y, z)\eta$ but multiplied by the differential energy shift $\eta = \frac{\Delta_1 - \Delta_2}{\Delta}$.

3.1 Atomic Motion in the Trap Without Collisions

Next, I would like to show that the fast oscillatory motion of the atoms can be averaged, and the detuning is then proportional to the energy of each atom. I consider atomic motion in a Gaussian trap without collisions. Since the temperature of the atoms is small compared to the maximum trap depth, I shall approximate the potential by an harmonic potential: $U(x, y, z) = \frac{m}{2}\left[\omega_x^2 x^2 + \omega_y^2 y^2 + \omega_z^2 z^2\right] - U_0$, where ω_i is the oscillation frequency of the ith axis. The motion of a specific atom is given then by $(x(t), y(t), z(t)) = (A_x \cos[\omega_x t + \phi_x], A_y \cos[\omega_y t + \phi_y], A_z \cos[\omega_z t + \phi_z])$, where A_i and ϕ_i correspond for the atom initial conditions. The accumulated relative phase between the two internal states is given by

Y. Sagi, *Collisional Narrowing and Dynamical Decoupling in a Dense Ensemble of Cold Atoms*, Springer Theses, DOI: 10.1007/978-3-642-29605-5_3, © Springer-Verlag Berlin Heidelberg 2012

$$\phi(t) = \int_0^t \Delta E(x(t'), y(t'), z(t'))dt'$$

$$\propto \frac{\eta}{8}[-8tU_0 + A_x^2 m\omega_x \left(2t\omega_x - \sin[2\phi_x] + \sin[2\phi_x + 2\omega_x t]\right)$$

$$+ A_y^2 m\omega_y \left(2t\omega_y - \sin[2\phi_y] + \sin[2\phi_y + 2\omega_y t]\right)$$

$$+ A_z^2 m\omega_z \left(2t\omega_z - \sin[2\phi_z] + \sin[2\phi_z + 2\omega_z t]\right)]. \qquad (3.1)$$

For times which are greater than the oscillation periods, $t \gg \omega_i^{-1}$, the two Sine terms for each axis can be neglect. Also, since there is an ensemble of atoms with different initial conditions (and therefore different values of ϕ_i), these sine terms will average out. We are then left with the expression

$$\phi(t) \propto -\eta t U_0 + \frac{\eta m}{4}\left[A_x^2 \omega_x^2 + A_y^2 \omega_y^2 + A_z^2 \omega_z^2\right]t = -\eta\left[U_0 + \frac{E_t}{2}\right]t, \qquad (3.2)$$

where I denoted by E_t the total energy of this atom. The result is that if the mean time between collisions is larger the the oscillation period in the trap, the fast oscillations can be averaged, and the rate of phase accumulation (detuning) depends on the average energy. For the experiments presented in this thesis the oscillation period is always the shortest timescale and therefore we can safely consider a coarse grained model in which the energy of each atom is constant in between collisions. When an elastic collision occurs, the total energy of each atom is changed, and therefore also its detuning.

3.2 The Hamiltonian and the Ensemble Coherence

Full quantum mechanical treatment of the ensemble will lead to a multi-particle Hamiltonian which is difficult to solve. Since interactions are small, I shall adopt instead a mean-field approach in which the ensemble in treated as an effective two-level system interacting with a modified external field which includes the effect of the atomic interactions. The effective single particle Hamiltonian for atoms with internal states designated by $|1\rangle$ and $|2\rangle$ can then be written as

$$\hat{H} = \hbar[\omega_0 + \delta(t)]|2\rangle\langle 2| + \hbar\Omega(t)|2\rangle\langle 1| + h.c., \qquad (3.3)$$

where ω_0 is the free-space transition frequency between the states, $\delta(t)$ is the detuning from the resonance due to the mean-field and $\Omega(t)$ is an external control field which is used for state manipulation (for the dynamical decoupling or Ramsey experiments). Here, the detuning is already assumed to be after the averaging over the fast oscillation has been performed. Starting from an initial state $|\psi(0)\rangle = 2^{-1/2}(|1\rangle + |2\rangle)$ and no external control fields, the wave-function at any given time is given in the rotating frame by $|\psi(t)\rangle = 2^{-1/2}(|1\rangle + e^{-i\phi(t)}|2\rangle)$, where the phase difference is given by $\phi(t) = \int_0^t \delta(t')dt'$. The ensemble coherence is quantified by the function $R(t) = \frac{|\langle\rho_{12}(t)\rangle|}{|\langle\rho_{12}(0)\rangle|}$, where ρ_{12} is the off-diagonal element of the reduced two-level density matrix and the notation $\langle\rangle$ stands for the ensemble average [1]. Due to the differential energy shifts of the trapping potential and to interactions between the atoms, each atom "sees" a different detuning. This detuning distribution generates a phase distribution with a width that increases in time and leads to dephasing—a reduction of the $R(t)$.

To get a feeling for the behavior of the coherence, we consider a Gaussian phase distribution, P_ϕ, with a standard deviation σ_ϕ. The calculation of the coherence is straight forward, giving

$$R(t) = e^{-\frac{\sigma_\phi^2(t)}{2}}. \tag{3.4}$$

This result shows that the coherence decays as the width of the phase distribution increases.

3.3 How to Measure Coherence?

The conventional way to measure coherence is in a Ramsey experiment [2]. Such a measurement consists of two short $\pi/2$ pulses separated in time, and a detection of the population at one of the states following. One way to perform this measurement is by keeping the time between the pulses constant and scanning the phase of the second pulse, which gives the coherence for that given waiting time. In this work, however, I use a slightly different sequence in which we keep the phase of both pulses the same, and scan the waiting time between the two pulses. In order to extract the coherence, we choose the detuning between the control field and the transition frequency to be large compared to the decoherence rate. This results in fast oscillations at the corresponding frequency, and we now explain why the envelope of these oscillations indeed gives the coherence as we have defined it above.

We consider a specific atom which is initially at state $|1\rangle$. The application of a $\pi/2$ pulse induce the following rotations:

$$|1\rangle \rightarrow \frac{1}{\sqrt{2}}(|1\rangle + |2\rangle)$$

$$|2\rangle \rightarrow \frac{1}{\sqrt{2}}(-|1\rangle + |2\rangle).$$

After the first Ramsey pulse the atom is transferred to the state $|\psi\rangle = \frac{1}{\sqrt{2}}(|1\rangle + |2\rangle)$. The atom now accumulates a relative phase between the two states, and at time t its state is given by $|\psi\rangle = \frac{1}{\sqrt{2}}(|1\rangle + e^{-i\phi(t)}|2\rangle)$. Applying the second $\pi/2$ Ramsey pulse transforms its state into $|\psi\rangle = \frac{1}{4}[(1 - e^{-i\phi})|1\rangle + (1 + e^{-i\phi})|2\rangle]$. We finally measure the population in state $|2\rangle$ and the result is given by $P_1 = |\langle 2| \psi\rangle|^2 = \frac{1}{2}(1 + \cos\phi)$. Since there are many atoms contributing incoherently to the signal, the experimental signal will be $P_1 = \frac{1}{2}(1 + \langle\cos\phi\rangle)$, where $\langle\rangle$ denote the ensemble average.

In order to obtain the envelope of the Ramsey signal one has to measure both signal quadratures and repeat the Ramsey sequence as before but set the phase of the second pulse to $\pi/2$ relative to the first pulse. The transformation of the second Ramsey pulse is then given by

$$|1\rangle \rightarrow \frac{1}{\sqrt{2}}\left(|1\rangle + i\,|2\rangle\right)$$

$$|2\rangle \rightarrow \frac{1}{\sqrt{2}}\left(i\,|1\rangle + |2\rangle\right),$$

and resulting experimental signal is $P_2 = \frac{1}{2}(1 + \langle\sin\phi\rangle)$. We normalize the Ramsey signal $2P - 1$ so they are centered at zero and spanning from -1 to 1, and their envelope is given by $R = \sqrt{(2P_1 - 1)^2 + (2P_2 - 1)^2} = \sqrt{\langle\cos\phi\rangle^2 + \langle\sin\phi\rangle^2}$. This can be also written as $R = |\langle\rho_{12}\rangle|$, since $\rho_{12} = e^{i\phi}$. This shows that by extracting the envelope of the Ramsey signal and normalizing it between 1 and 0, one can measure the coherence. Instead of measuring both quadratures in the experiment, it is enough to set the detuning of the control field to a large enough value and directly extract the envelope, as was explained in Chap. 2.

3.4 Dephasing in a 3D Harmonic Trap Without Collisions

For atoms trapped in a 3D harmonic potential the phase distribution is not Gaussian. Follow Ref. [3], we write the phase distribution as: $P_\phi = \frac{3\sqrt{3}\left(-\mu_\phi + \sqrt{3}\sigma_\phi + \phi\right)^2}{2\sigma_\phi^3}$ $e^{-\frac{\sqrt{3}\left(-\mu_\phi + \sqrt{3}\sigma_\phi + \phi\right)}{\sigma_\phi}}$, where μ_ϕ and σ_ϕ are its average and standard deviation, respectively. Using this equation we can calculate the coherence $R = \left[1 + \frac{\sigma_\phi^2}{3}\right]^{-3/2}$. Without collisions the time dependence is given by $\sigma_\phi(t) = \sigma_\delta t$, and inserting this into the previous expression we obtain $R(t) = \left[1 + \frac{(\sigma_\delta t)^2}{3}\right]^{-3/2}$. Solving for τ_1 the equation $R(\tau_1) = e^{-1}$ yields: $\tau_1 = \sqrt{3\left(e^{2/3} - 1\right)}\sigma_\delta^{-1} \approx 1.69 \cdot \sigma_\delta^{-1}$. Measuring the coherence in experiments with very low densities we extract τ_1, from which we can directly calculate σ_δ. It can also be shown that σ_δ is linearly proportional to the temperature T [3], and therefore we can write $\tau_1 = \eta T^{-1}$. We have carried out experiments at very low density and measured $\eta = 1.2 \cdot 10^{-7}$ K$\cdot$s. This value differs by a factor of 2 from the value derived in Ref. [3] due to a reason yet unknown.

3.5 The Ensemble Coherence with an External Control Field: A System-Reservoir Gaussian Framework

We now consider the evolution of the ensemble coherence in the presence of external control field. A specific useful form of external control is the π-pulse—an external field pulse with a duration and power chosen such that it induces a complete population inverting between the internal states of the atoms. Assuming a Gaussian phase distribution, the coherence with a series of π-pulses is given by [1, 4]:

$$R(t) = e^{-\int_0^\infty d\omega S_\delta(\omega) F(\omega t)/\pi\omega^2}, \tag{3.5}$$

where the argument is the overlap integral between the fluctuations power spectrum, $S_\delta(\omega) = \int_{-\infty}^\infty \langle \delta(t)\delta(0)\rangle e^{i\omega t} dt$, and a filter function which encapsulates the information on the dynamical decoupling pulse sequence and is given by $F(\omega t) = \frac{1}{2}|\sum_{k=0}^n (-1)^k (e^{i\omega t_{k+1}} - e^{i\omega t_k})|^2$ with t_k being the pulses times, $t_0 = 0$ and $t_{n+1} = t$. Since collisions form a Poisson process, we expect the detunings correlation function to decay exponentially. The power spectrum is therefore a Lorentzian $S_\delta(\omega) = \int_{-\infty}^\infty \Phi_\delta(t) e^{i\omega t} dt = \frac{2\Gamma\sigma_\delta^2}{\Gamma^2+\omega^2}$.

References

1. Cywiński L, Lutchyn RM, Nave CP, Sarma SD (2008) Phys. Rev. B 77:174509
2. Ramsey NF (1956) Molecular beams. Oxford University Press, Oxford
3. Kuhr S, Alt W, Schrader D, Dotsenko I, Miroshnychenko Y, Rauschenbeutel A, Meschede D (2005) Phys. Rev. A 72:023406
4. Gordon G, Kurizki G, Lidar DA (2008) Phys. Rev. Lett. 101:010403

Chapter 4
Spectral Narrowing due to Elastic Collisions

4.1 Introduction to Motional Narrowing

As already explained, inhomogeneities in δ over the ensemble lead to dephasing of a stored coherence. Intriguingly, fluctuations in δ can prolong the coherence time—a phenomenon called motional narrowing. Historically, motional narrowing was first observed in liquid NMR, where a large reduction in the width of spectral lines was observed in comparison to solid NMR due to the thermal motion of the nuclei [1]. Later the effect was also reported in other fields such as molecular physics [2], semiconductor microcavities [3] and quantum dots [4]. In this chapter we study motional narrowing due to elastic collisions in a dense atomic ensemble. In contrast to previous experiments, our apparatus enables precise and independent control over the thermodynamic parameters of the system. Owing to this, we are able to quantitatively analyze the dependance of the narrowed linewidth on the fluctuations rate and strength, and demonstrate an inverse linear dependence on the former and quadratic dependence on the latter. We also find that the narrowed linewidth exhibits universal scaling with the atomic phase space density.

4.2 Collisional Narrowing: Motional Narrowing due to Elastic Collisions

The general idea behind collisional narrowing in cold atomic ensembles is that due to velocity-changing collisions the evolution of the phase distribution transform from ballistic expansion to diffusion, which results in a much slower decoherence. In the limit where collisions are not important, namely $t \ll t_{col}$, the phase distribution is ballistically expanding, $\sigma_\phi = \sigma_\delta t$, and using Eq. (3.4) we get that the coherence is given by

$$R_{t \ll t_{col}}(t) = e^{-t^2/\tau_1^2}, \tag{4.1}$$

Y. Sagi, *Collisional Narrowing and Dynamical Decoupling in a Dense Ensemble of Cold Atoms*, Springer Theses, DOI: 10.1007/978-3-642-29605-5_4,
© Springer-Verlag Berlin Heidelberg 2012

where $\tau_1 = \sqrt{2}\sigma_\delta^{-1}$, and it corresponds to a Gaussian lineshape. Note also that when the detuning distribution is not a Gaussian (Eq. 4.1) is only an approximation with $\tau_1 = \alpha\sigma_\delta^{-1}$, where α a number on the order of 1 which can be derived from the distribution (for 3D harmonic trap it is $\alpha \approx 1.69$).

In the other limit, when collisions change the dynamics of the coherence, the accumulated phase is given by $\phi(t) = \Sigma_{i=1}^{N}\Delta\phi_i = \Sigma_{i=1}^{N}t_i\delta_i$, where $\{\delta_i\}$ are drawn from P_δ, and $\{t_i\}$ is a series of durations between collisions, which are distributed exponentially with a rate constant Γ_{col}. Note that Γ_{col}^{-1} is the energy autocorrelation time constant, which for cold atoms in $3D$ relates to the conventionally defined collision rate $\tilde{\Gamma}_{col}$ by $\Gamma_{col} = \tilde{\Gamma}_{col}/2.7$ [5]. Employing the central limit theorem, we obtain that $\phi(t)$ is normally distributed with a mean $\mu_\phi = \mu_\delta t$ and a standard deviation $\sigma_\phi = \sqrt{D \cdot t}$, where D is a diffusion coefficient given by $D = 2t_{col}\sigma_\delta^2$. We plug this result into (Eq. 3.4) and get that the decay envelope is given in this limit by

$$R_{t \gg t_{col}}(t) = e^{-t/\tau_2}, \tag{4.2}$$

$$\text{where } \tau_2 = 2D^{-1} = \Gamma_{col}\sigma_\delta^{-2} = \alpha^{-2}\Gamma_{col}\tau_1^2. \tag{4.3}$$

The exponential decay is the hallmark of motional narrowing resulting in a Lorentzian-shaped transition line where the width is inversely proportional to the collision rate. Note that since we have employed the central limit theorem, this result is only valid for detuning distributions with a finite second moment. In Chap. 6 (Eq. 6.7) we obtain a generalization of this expression which is valid also for heavy-tailed distributions.

Using the general expression for a Gaussian phase distribution given in (Eq. 3.5) with the time-domain Ramsey experiment filter function, $F(z) = 2\sin^2\frac{z}{2}$, and with the Lorentzian power spectrum, $S_\delta(\omega) = \frac{2\Gamma_{col}\sigma_\delta^2}{\omega^2+\Gamma_{col}^2}$, the coherence is found to be a generalized Gumbel function:

$$R_{Gumbel}(t) = e^{-\sigma_\delta^2\Gamma_{col}^{-2}\left(e^{-\Gamma_{col}t}+\Gamma_{col}t-1\right)}, \tag{4.4}$$

which converge to the functional forms discussed before in the appropriate limits. Note that this function is only valid for $\delta(t)$ which are a *Gaussian process*. In particular, for atoms in a $3D$ harmonic trap the exact detuning distribution is not a Gaussian, and therefore this function is only an approximation, albeit a very good one. In (Sect. 5) we solve exactly a discrete fluctuation model which does not assume a specific functional form for the detuning distribution. The disadvantage of the exact spectrum, though, is that it is not given explicitly but only numerically in terms of the system parameter (i.e. σ_δ and Γ_{col}). This makes this approach inconvenient for data analysis and therefore we use the exact spectrum only as a mean to correct the systematic deviations from the Gumbel function given in Eq. (4.4). For more details on the data analysis see Appendix A.

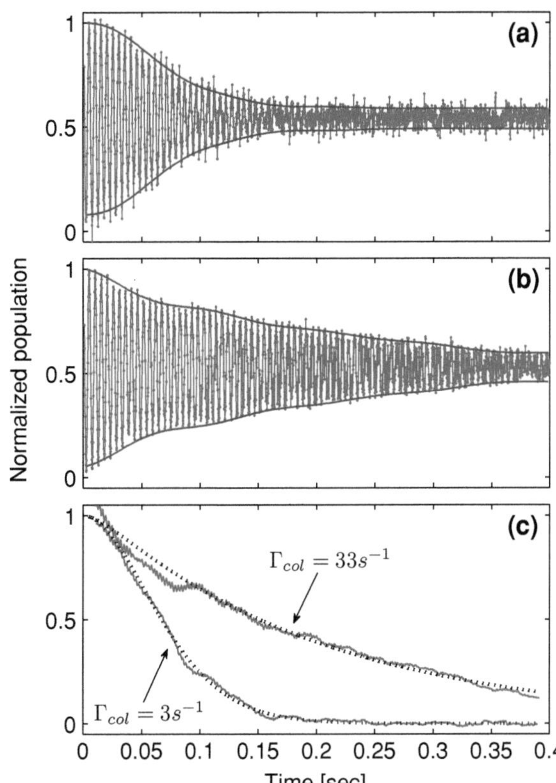

Fig. 4.1 Typical results of Ramsey experiments with cold ^{87}Rb atoms. Two short $\pi/2$ pulses are given, separated by a time indicated by the x-axis. The y-axis is the normalized population at $|2\rangle$. The data presented here was taken for atoms with temperature of $1.7\,\mu$K, which gives a dephasing time of $\tau_1 = 73$ ms. The collision rates are (**a**) $\Gamma_{col} = 3s^{-1}$ (**b**) $\Gamma_{col} = 33\,s^{-1}$, and graph (**c**) is a comparison of the envelopes of the two experiments, normalized to begin at 1 (*solid blue line*). The dotted lines are fits to a Gumbel function, as defined by Eq. (4.4). The $\pi/2$ pulse duration is $\sim$300 μs. The detuning of the control field is $2\pi \cdot 203$ Hz, chosen such that the envelope can be easily extracted. The envelopes are extracted by calculating the standard deviation of all points in a single Ramsey oscillation and multiplying by $\sqrt{2}$

4.3 Experimental Observation of Collisional Narrowing

The typical temperature in the experiment is $T = 2\,\mu$K, low enough that we can approximate our Gaussian trap by an harmonic potential, and the density is $\rho = 10^{13}$ cm^{-3}. The atomic phase space density is smaller than 0.05 which means that the motion of the atoms in the trap can be treated classically. Typical results of two Ramsey experiments at low and high collision rates are depicted in Fig. 4.1. For $t < \Gamma_{col}^{-1}$ the two envelopes have the same Gaussian like decay shape as expected by Eq. (4.1), but for longer times the envelope of the Ramsey experiment with the higher collision rate deviates and changes its form to an exponential like with a lower dephasing rate.

We fit the envelopes extracted from the Ramsey measurements with a Gumbel function (see Fig. 4.1c). We have carried out three sets of experiments with different temperatures, in each of which we have varied the density and extracted τ_2. The measured temperature is used to calculate τ_1 for each datasets, and the extracted τ_2 is corrected for the non-Gaussian detuning distribution of atoms in a 3D harmonic trap (for more details see Appendix A). In Fig. 4.2a we plot the extracted values

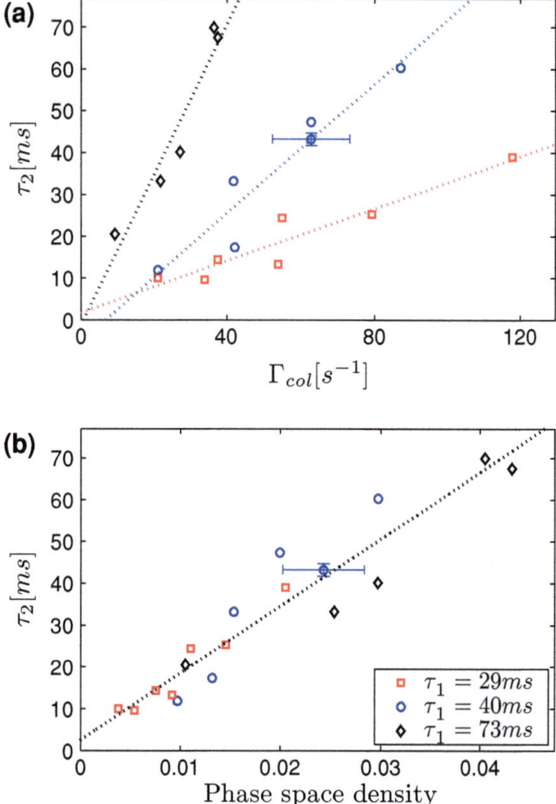

Fig. 4.2 The exponential decay time τ_2 as a function of the measured collision rate (**a**) and phase space density (**b**) for three datasets with different temperature and τ_1. The collision rate is the average collision rate in the cloud and it is calculated from the measured density, temperature and oscillation frequency of the trap. *The dotted lines are linear fits to the data.* The measured temperature is 1.7 μK (*diamonds*), 3.1 μK (*circles*) and 4.3 μK (*squares*)

of τ_2 for the three datasets, and as predicted by Eq. (4.3), we get that it depends linearly on the collision rate with a different slope for each temperature. An important consistency test is the comparison of the value of τ_1 calculated from the slopes of the linear fits of Fig. 4.2a to the value calculated directly from the temperature. We use the relation for a 3D harmonic trap $\sigma_\delta^{-1} = 1.69 \cdot \tau_1$ and Eq. (4.3) and find $\tau_1 = 71 \pm 18$ ms, 47 ± 10 ms and 30 ± 6 ms for the datasets with temperatures of 1.7 μK, 3.1 μK and 4.3 μK, respectively. These values of τ_1 are in good agreement with the values $\tau_1 = 73$ ms, 40 ms and 29 ms, calculated from the measured temperatures. The origin of axis is within the error margins of the three linear fits.

4.4 Universal Behavior of the Narrowed Linewidth

A striking universal behavior is revealed when Eq. (4.3) is rewritten in terms of the system thermodynamic parameters. The density of atoms can be written $\rho \sim \Phi T^{3/2}$, where Φ is the phase space density, and T is the temperature. The collision rate scaling is $\tilde{\Gamma}_{col} = \rho \sigma_{col} v_{th}$ where σ_{col} is the collision cross-section, and v_{th} is the average thermal velocity which is proportional to $T^{1/2}$. For low temperatures the collisions are s-wave scattering processes and σ_{col} does not depend on density or temperature. The scaling of the collision rate is therefore $\Gamma_{col} \sim \rho T^{1/2}$, and when substituted into Eq. (4.3) we get $\tau_2 \sim \Phi T^2 \sigma_\delta^{-2}$. It can be shown that $\sigma_\delta \sim T$ for any potential of the form $U \sim x^n$, and in particular this is the case for an harmonic potential. The final result is that the narrowed linewidth is inversely proportional to the atomic phase space density: $\tau_2 \sim \Phi$. In Fig. 4.2b we plot τ_2 of the same three experimental datasets presented before versus the measured phase space density. As predicted, all data points lie on a linear curve. We fit the data with a power law function $\tau_2 = C \cdot \Phi^n$ and find $n = 0.91 \pm 0.19$, in agreement with the expected value of $n = 1$. A linear fit yields a slope of $\tau_2/\Phi = 1602 \pm 285\,ms$ in agreement with the calculated value of $1629\,ms$ based on the trap parameters. All error margins are given for a 95 % confidence level.

4.5 Monte-Carlo Numerical Simulations

To further support our findings we perform molecular dynamics Monte-Carlo simulations. We simulate classical motion of 4000 atoms in a cigar shaped 3D harmonic trap with parameters similar to the experiment. The atoms are drawn from a Boltzmann distribution assuming a temperature of $4\,\mu K$. The collisions are simulated in the following way: we use the steady state density profile to calculate each atom's probability to undergo a collision according to the collision rate at its position. For each atom undergoing a collision, we calculate its velocity after the collision by assuming a virtual counterpart with a velocity which is drawn according to a probability distribution which depends on the velocity of the first atom, and assuming s-wave scattering processes. Finally, We calculate the energy shift of the internal states induced by the external potential along the trajectory of each atom, and integrate this to get the accumulated phase and the contribution to the Ramsey signal. Using τ_1 as the scaling parameter of time, we rewrite Eq. (4.3) in a dimensionless form: $\tau_2/\tau_1 = \alpha^2 \cdot \Gamma_{col} \tau_1$. In Fig. 4.3 we plot the experimental results in a dimensionless form, and find that they agree well both with the theory and Monte-Carlo simulations.

4.6 Collisional Narrowing in a Symmetric Many-Body Coherent Superposition

In our experiment the atomic ensemble is treated as an effective single spin system, which also correctly describes an atomic memory based on a collinear pump-probe EIT configuration [6]. Other schemes for creating non-classical states of light are

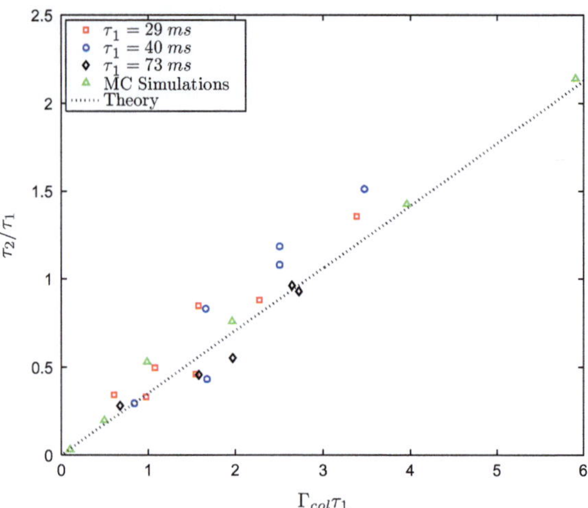

Fig. 4.3 A comparison of the dimensionless collisional narrowing timescale, τ_2/τ_1, versus the dimensionless collision rate $\Gamma_{col}\tau_1$ for experimental data (*red squares*, *blue circles* and *black diamonds*), Monte-carlo simulations (*green triangles*) and the theoretical prediction of Eq. (4.3) (*black dotted line*)

based on Raman scattering processes and they induce a global coherence in the ensemble [7, 8]. As an example, in a co-linear geometry between the Raman read and write beams the state of the ensemble can be written as $|\psi(0)\rangle = \frac{1}{\sqrt{N}} \sum_{k=1}^{N} a_k^\dagger |0\rangle$, where the operator $a_k^\dagger$ creates an excitation in the k-th atom and N is the total number of atoms. The ensemble state at a time t is given by $|\psi(t)\rangle = \frac{1}{\sqrt{N}} \sum_{k=1}^{N} e^{-i\phi_k(t)} a_k^\dagger |0\rangle$, where $\phi_k(t) = \int_0^t dt\, \delta_k(t)$ and $\delta_k(t)$ is the detuning realization of the k-atom. The fidelity is given by $\mathcal{F} = |\langle\psi(t)|\psi(0)\rangle| = |\frac{\sum_{k=1}^{N} e^{-i\phi_k(t)}}{N}|$. Approximating this sum by the integral $\mathcal{F} = |\int_{-\infty}^{\infty} d\phi\, P_\phi(t) e^{-i\phi(t)}|$ and using a Gaussian phase distribution yields $\mathcal{F} = e^{-\frac{\sigma_\phi^2(t)}{2}}$—identical to the decay envelope in Eq. (3.4). This shows that as long as the atomic trajectories are not affected by the global coherence, the effect of collisions on the coherence time in a Raman scattering scheme is the same as discussed before.

References

1. Bloembergen N, Purcell EM, Pound RV (1948) Phys Rev 73:679
2. Oxtoby DW (1979) J Chem Phys 70:2605
3. Whittaker DM, Kinsler P, Fisher TA, Skolnick MS, Armitage A, Afshar AM, Sturge MD, Roberts JS (1996) Phys Rev Lett 77:4792
4. Berthelot A, Favero I, Cassabois G, Voisin C, Delalande C, Roussignol P, Ferreira R, Gerard JM (2006) Nat Phys 2:759
5. Monroe CR, Cornell EA, Sackett CA, Myatt CJ, Wieman CE (1993) Phys Rev Lett 70:414
6. Lukin MD (2003) Rev Mod Phys 75:457
7. Duan LM, Lukin MD, Cirac JI, Zoller P (2001) Nature 414:413
8. Kuzmich A, Bowen WP, Boozer AD, Boca A, Chou CW, Duan LM, Kimble HJ (2003) Nature 423:731

Chapter 5
The Ensemble Spectrum with an Arbitrary Detuning Distribution

In this chapter we aim to exactly solve for the spectrum (and the decay of coherence) of an ensemble with an arbitrary detuning distribution $P_0(\delta)$. Historically, the spectrum of a TLS ensemble was first studied by Kubo who considered an ensemble of harmonic oscillators with randomly varying resonant frequencies, and found an analytic expression for the ensemble spectrum for a frequency which is a Gaussian process [1]. In the prevalent case, where the frequency autocorrelation function decays exponentially in time, the Kubo solution gives a generalized Gumbel function, as defined in Eq. (4.4).

5.1 Discrete Fluctuations Model and the Resulting Spectrum

In order to obtain an analytic solution for any detuning distribution, we consider fluctuations in the detuning which are discrete in nature and follow Poisson statistics (see upper schematics in Fig. 5.1). A new detuning is drawn from $P_0(\delta)$ for any TLS undergoing a randomizing event, which occurs at a rate Γ. We assume that two consecutive detunings have no correlation, which together with the Poissonian statistics of the randomizing events yields the correlation function $C(|t - t'|) = \langle \delta(t)\delta(t') \rangle = e^{-\Gamma|t-t'|}$. The ensemble coherence for this model is given by [2]

$$\tilde{R}(s) = \frac{\tilde{R}_0(s + \Gamma)}{1 - \Gamma \tilde{R}_0(s + \Gamma)}, \tag{5.1}$$

where $\tilde{R}(s) \equiv \mathcal{L}\{R(t)\}$ is the Laplace transform of $R(t)$, and $R_0(t)$ is the coherence without fluctuations written explicitly as

$$R_0(t) = \int_{-\infty}^{\infty} P_0(\delta)e^{i\delta t} d\delta. \tag{5.2}$$

Y. Sagi, *Collisional Narrowing and Dynamical Decoupling in a Dense Ensemble of Cold Atoms*, Springer Theses, DOI: 10.1007/978-3-642-29605-5_5, © Springer-Verlag Berlin Heidelberg 2012

Knowing $\tilde{R}(s)$ we can calculate the complex spectrum of the TLS ensemble by $S(\omega) = R(i\omega) + R(-i\omega)$ and the decay of coherence $R(t) = \mathcal{L}^{-1}\{\tilde{R}(s)\}$. For an alternative original derivation of Eq. (5.1) see Appendix B.

5.2 Comparison with a Gaussian Process

We compare the result of Eq. (5.1) with the predictions of the Kubo model. As mentioned before, Kubo showed that for a $\delta(t)$ which is a *Gaussian process* with an exponentially decaying correlation function the coherence is given by a generalized Gumbel function [1]

$$R_{Gumbel}(t) = e^{-\sigma_\delta^2 \Gamma^{-2}\left(e^{-\Gamma t}+\Gamma t-1\right)}, \tag{5.3}$$

where σ_δ and Γ^{-1} are the standard deviation of $\delta(t)$ and its correlation time, respectively. In the limit of $\Gamma \to 0$ one retrieves a Gaussian inhomogeneous coherence decay: $R_0(t) = e^{-t^2/\tau^2}$, with $\tau = \sqrt{2}\sigma_\delta^{-1}$, a result which can be also obtained directly from Eq. (5.2) and using a Gaussian phase distribution $P_0(\delta) = (2\pi\sigma_\delta^2)^{-1/2}e^{-\delta^2/2\sigma_\delta^2}$. Intuitively, one would expect to retrieve the generalized Gumbel function by using the solution of the discrete model given in Eq. (5.1) with this $R_0(t)$. In Fig. 5.1 we compare the two spectra and find that they differ. This is rather surprising since the detuning distribution $P_0(\delta)$ and correlation function $C(t)$ are *identical* for the two models. The difference stems from the discrete nature of the frequency fluctuations which results in an accumulated phase not distributed as a Gaussian. In other words, the discrete fluctuation model give rise to $\delta(t)$ which is not a *Gaussian process*, even when the underlying detuning distribution is Gaussian.

To gain better understanding of this point we have carried out numerical simulations with 20, 000 TLS, for each of which the detunings are drawn from a Gaussian distribution and the times between collisions are drawn from an exponential distribution with a rate Γ. The results of the simulation are shown as squares in Fig. 5.1, and fully conform to the result of the discrete model. In a second simulation the rate of randomizing events was ten times higher, but the detunings were drawn with correlations to each other according to $\sigma_\delta^{-2}\langle\delta(t)\delta(0)\rangle = e^{-t/t_{corr}}$, with a time-constant $t_{corr} = 10\Gamma^{-1}$ (for more details see Appendix C). The results of this simulation are in good agreement with the spectrum of the generalized Gumbel function, as can be seen in Fig. 5.1. The correlations between subsequent detunings effectively smooth the spectral jumps, and therefore increasing $t_{corr}\Gamma$ gradually transforms $\delta(t)$ to be a Gaussian process. The quantity $t_{corr}\Gamma$ measures the "hardness" of the fluctuations, with a "soft" Kubo solution obtained for $t_{corr}\Gamma \gg 1$.

The spectral sensitivity can be used as a tool to study the underlying physics of the fluctuations. As an example we consider a vapor cell in a room temperature with two species of atoms; active atoms at which the coherence is going to be stored and neutral atoms which are usually refereed to as a "buffer gas", added to slow

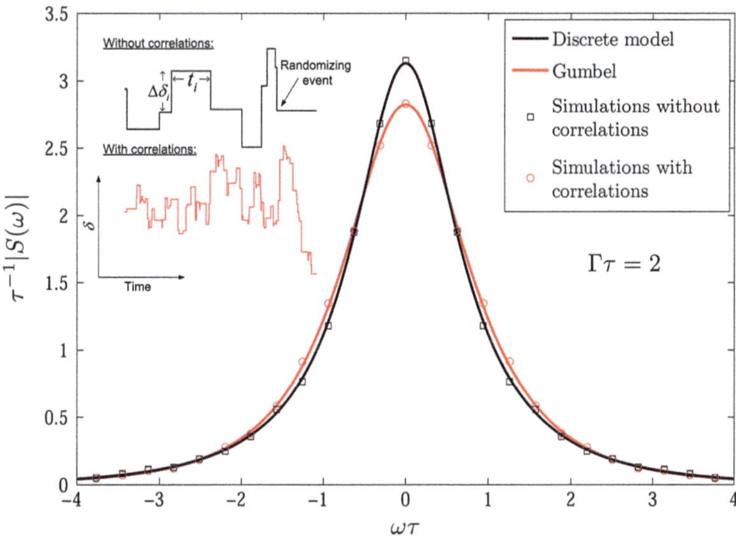

Fig. 5.1 The spectrum of an ensemble with a Gaussian detuning distribution as calculated by the solution of the discrete fluctuation model given in Eq. (5.1) (*solid black line*), and by the Kubo's solution for a Gaussian process given in Eq. (5.3) (*red-dotted line*). The markers are the results of simulations without correlations between subsequent detunings (*black squares* and *upper part* of the schematics), and with correlations (*red circles* and *lower part* of the schematics)

the diffusion process of the active atoms which blur the spatial information. Due to the relatively high temperature of the vapor cells, collisions in this ensemble are a many-body multi-channel scattering process which is not well understood [3]. The spectrum of the atoms, which is initially a Gaussian due to Doppler broadening, is narrowed because of the collisions—a phenomenon called "Dicke narrowing" [4]. Though the original phenomenon considered by Dicke was for a single photon transition, in hot vapor cells it is usually seen in two-photon EIT measurements [5, 6]. In Fig. 5.2 the ratio of the spectral width of a soft Kubo-like ensemble and a hard discrete-like ensemble is plotted for the Gaussian case. The calculation shows that the difference can be as large as 18 %, a value which is most likely measurable in an experiment. A measurement of the Dicke narrowed spectrum in such an ensemble can therefore quantitatively distinguish between soft and hard collisions. It is expected that whether the collisions are soft or hard will be determined by the mass ratio of the active atom and buffer gas atom. It will be, therefore, interesting to perform this experiment with different kind of buffer gas. Of course, such a measurement requires an independent determination of Γ and τ. Also, to ensure the difference is maximal, one has to tune the parameters such that $\Gamma\tau$ is on the order of 1. This can be achieved by changing the angle between the pump and the probe beams, and by controlling the pressure of the buffer gas.

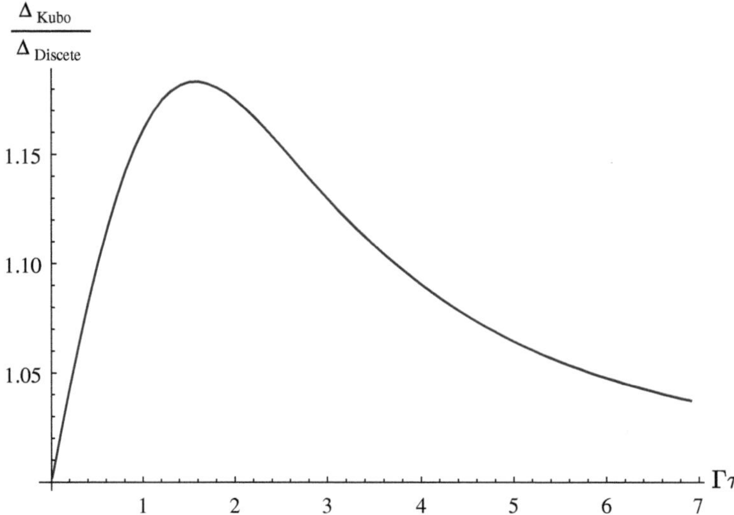

Fig. 5.2 The ratio of the full width at half maximum (FWHM) of the Kubo spectrum and the discrete fluctuations model spectrum, as a function of $\Gamma\tau$, where τ is the inhomogeneous decay time. The difference between the two FWHM grows as high as 18 % for $\Gamma\tau \approx 1.5$, a value which is most likely measurable in an experiment

5.3 Application of the Discrete Fluctuations Model to Cold Atomic Ensembles

A wide-spread physical realization of the discrete model consists of an ensembles of cold atoms trapped in a conservative potential. The detuning distribution which originates from differential shift (e.g. differential light or Zeeman shifts) is determined by the geometry and dimensionality of the trapping potentials and by the atomic temperature [7]. The source of randomization are elastic collisions (s-wave scattering) which are predominant at low temperatures. Every 2.7 collisions, on average, the energy of each atom is randomized [8], and as a consequence also its potential energy and detuning. The detuning distribution for a 3D harmonic trap is proportional to the density of states and the Boltzmann factor: $P_0(\delta) \sim \delta^2 e^{-K\delta}$, with $K = 2\hbar/\eta k_B T$ and η is a dimensionless parameter characterizing the differential shifts in the trap [7]. The resulting inhomogeneous decay is $R_0(t) = \left(1 + \frac{t^2}{\tau_*^2}\right)^{-\frac{3}{2}}$ with $\tau_* = \tau/\sqrt{e^{2/3} - 1}$. The Laplace transform of $R_0(t)$ is given by

$$\tilde{R}_0(s) = \tau_*^2 s \left[\frac{\pi}{2} H\left(1, \tau_* s\right) - \frac{\pi}{2} Y\left(1, \tau_* s\right) - 1\right], \tag{5.4}$$

where H is the Struve function and Y is the Bessel function of the second kind. We use Eq. (5.1) to calculate $\tilde{R}(s)$, from which we compute the inverse Laplace

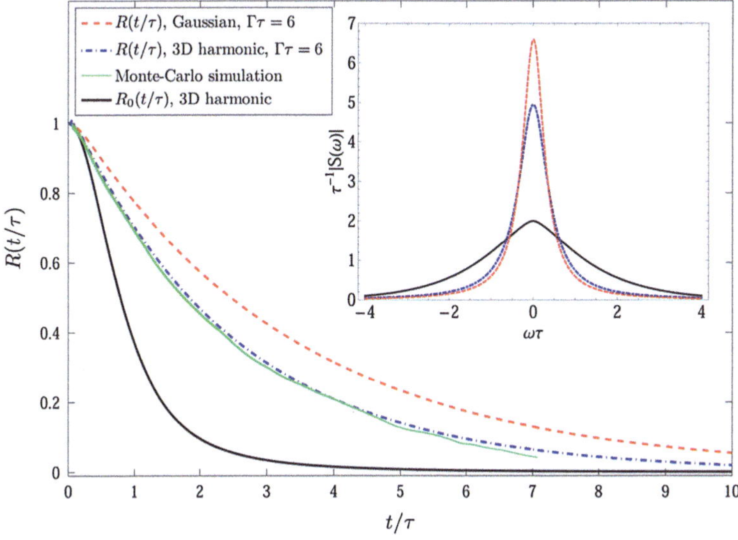

Fig. 5.3 A plot of the decay of coherence in the absence of fluctuations, $R_0(t/\tau)$, for a 3D harmonic trap (*black solid line*) compared to the decay with fluctuations, $R(t/\tau)$, with $\Gamma/\tau = 6$ calculated by the solution of the discrete model for a 3D harmonic trap (*blue dashed-dotted line*) and for a Gaussian detuning distribution (*red dashed line*). The inset shows the corresponding spectra. Also plotted is the result of a Monte-Carlo simulation with the same Γ and τ (*green solid line*). The simulations were done with 4000 atoms and colliding at an average rate $\Gamma_{col} = 2.7\Gamma$ [8]

transform numerically and obtain $R(t)$. In Fig. 5.3 we plot $R_0(t/\tau)$ and $R(t/\tau)$ for $\Gamma\tau = 6$ and compare it to the solution of the discrete model with a Gaussian detuning distribution. For both the Gaussian and the 3D harmonic trap cases with $\Gamma\tau > 0$ the decay is slower than for a $R_0(t/\tau)$—the well-known motional narrowing effect [9, 10]. However, the difference between the exact solution for 3D harmonic trap and an approximated Gaussian solution is larger than 25 % at the $1/e$ decay point and 65 % at $1/e^2$. This shows that even for the same values of τ and Γ the dependence on the detuning distribution $P_0(\delta)$ is significant. We further corroborated the discrete model for trapped atomic ensemble by performing 3D Monte-Carlo (MC) simulations of atoms trapped by a far-off-resonance laser, in the same way as explained before. The results of the simulations are plotted in Fig. 5.3. The discrete model solution for a 3D harmonic trap agrees well with the MC simulations, whereas the Gaussian solution deviates considerably from both. We conclude that the discrete model is better suited to describe the spectrum of cold atomic ensembles.

5.4 Comparison of the Theory to Experimental Results

The experiments are performed with a trap with radial oscillation frequency of $\omega_r = 2\pi \cdot 520\,\text{Hz}$, which is measured directly by parametric excitation. The final laser power after the evaporation is $1.6\,\text{W}$. The thermodynamic parameters are measured after a time of flight with an absorption imaging technique, and typically yield $300{,}000$ atoms in a temperature of $3\mu\text{K}$ and a phase space density of ~ 0.05. The spin relaxation time (T_1) which originates from inelastic collisions [11] is measured to be $T_1 = 6\,\text{s}$. Since the total population in the states $|1\rangle$ and $|2\rangle$ is 70% in these experiments, we normalize the signal to the initial population at state $|2\rangle$ and also subtract the measured background noise so the resultant signal is normalized between 0 and 1. To facilitate the extraction of the Ramsey envelope, we set the external control field frequency such that the average detuning relative to ω_0 is much larger than $2\pi\tau^{-1}$. The extracted envelopes of the measured Ramsey fringes are depicted in Fig. 5.4 for $\Gamma = 130\,\text{s}^{-1}$ and 11s^{-1}.

We measure Γ directly by creating a sudden small perturbation in the trapping laser intensity and measuring the decay of the breathing-mode oscillations, as can be seen in the inset of Fig. 5.4. This value is also compared to indirect calculation using the measured temperature and density, and agrees to within 20%. By reducing the MOT beams intensity we change Γ with almost no change in the temperature. Thus, we can reduce Γ such that $\Gamma\tau \ll 1$ and fit the measured Ramsey signal with the expected $R_0(t)$ and extract τ. The theoretical decay curves, which are calculated without fitting parameters using Eqs. (5.1) and (5.4) and the measured Γ and τ, are plotted in Fig. 5.4) and agree well with the experimental data.

The asymptotic long time behavior of Eq. (5.1) for $t/\tau \gg 1$ should coincide with the results already obtained in Eqs. (4.2) and (4.3). For simplicity we assume that the envelope $R_0(t)$ is a positive monotonically decreasing function, which means that $\tilde{R}_0(s)$ is also a monotonically decreasing. The coherence given in Eq. (5.1) has a single pole at s_0 which is the solution of the equation: $1 - \Gamma\tilde{R}_0(s_0 + \Gamma) = 0$, and the long time limit behavior of $R(t)$ is then given by $R(t) = e^{-t/T_2}$, where $T_2 = -s_0^{-1}$. From dimensional considerations $R_0(t)$ can always be written as $R_0(t) = \Lambda(t/\tau)$. Using this notation the Laplace transform can be written as $\tilde{R}_0(s) = \tau\tilde{\Lambda}(\tau s)$, where $\tilde{\Lambda}(\tau s) = \int_0^\infty dx\,\Lambda(x)e^{-x\tau s}$. We use a Taylor expansion to first order around $s = 0$, $\tilde{R}_0(s + \Gamma) \approx \tau\tilde{\Lambda}(\Gamma\tau) + \tau^2\tilde{\Lambda}'(\Gamma\tau)s$, and solve for s_0:

$$T_2 = \Gamma\tau^2 \mathcal{F}(\Gamma\tau), \tag{5.5}$$

where $\mathcal{F}(x) = -\dfrac{\tilde{\Lambda}'(x)}{1 - x\tilde{\Lambda}(x)}$. Eq. (5.5) gives a closed-form expression for the motional-narrowing timescale, T_2, in terms our two model parameters τ and Γ for slowly varying $\mathcal{F}(\Gamma\tau)$. It can be shown that $\mathcal{F}(x) = 1/2$ for $\Lambda(x) = e^{-x^2}$, and $\lim_{x\to\infty}\mathcal{F}(x) = 1/3$ for $\Lambda(x) = (1 + x^2)^{-3/2}$, which corresponds to $T_2 = \Gamma\tau^2/2$ for the Gaussian case, and $T_2 = \Gamma\tau^2/3$ for 3D harmonic trap. This coincide with the result of Eq. (4.3). Using the last equation with the experimentally measured $\tau = 42\,\text{ms}$ and $\Gamma = 130\,\text{s}^{-1}$ we find $T_2 = 76.4\,\text{ms}$. We use the tail ($t > 130\,\text{ms}$) of

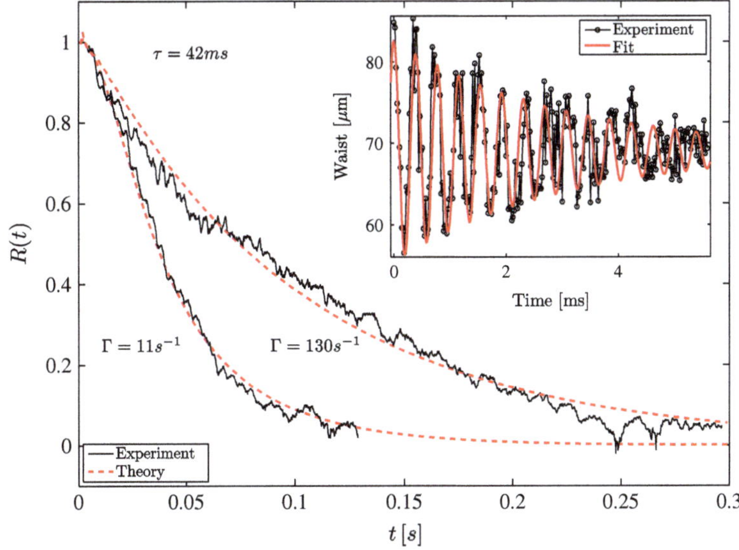

Fig. 5.4 The envelopes, $R(t)$, of two Ramsey experiments with the same inhomogeneous timescale $\tau = 42$ms and different Γ (*black solid lines*), and theoretical calculations using Eq. (5.4) and Eq. (5.1) without fitting parameters (*red dashed line*). The inset shows the decay of atomic-cloud waist oscillations fitted by an exponentially decaying cosine $e^{-\Gamma t}\cos(2\pi f_0 t)+c$, with $\Gamma = 130s^{-1}$, and f_0 which agrees well with the value measured by parametric excitation $f_0 = \omega_r/\pi$

the experimentally measured coherence decay curve (Fig. 5.4) and directly extract $\tau_2 = 76.3 \pm 1.2$ ms with a 95 % confidence level.

References

1. Kubo R (1961) Fluctuation, relaxation and resonance in magnetic systems. Scottish universities summer school in physics, Oliver and Boyd, Edinburgh, pp 23–68
2. Brissaud A, Frisch UJ (1974) Math Phys 15:524
3. Camparo J (2007) Phys Today 60:33
4. Dicke RH (1953) Phys Rev 89:472
5. Firstenberg O, Shuker M, Ben-Kish A, Fredkin DR, Davidson N, Ron A (2007) Phys Rev A 76:013818
6. Shuker M, Firstenberg O, Pugatch R, Ben-Kish A, Ron A, Davidson N (2007) Phys Rev A 76:023813
7. Kuhr S, Alt W, Schrader D, Dotsenko I, Miroshnychenko Y, Rauschenbeutel A, Meschede D (2005) Phys Rev A 72:023406
8. Monroe CR, Cornell EA, Sackett CA, Myatt CJ, Wieman CE (1993) Phys Rev Lett 70:414
9. Bloembergen N, Purcell EM, Pound RV (1948) Phys Rev 73:679
10. Berthelot A, Favero I, Cassabois G, Voisin C, Delalande C, Roussignol P, Ferreira R, Gerard JM (2006) Nat Phys 2:759
11. Widera A, Gerbier F, Folling S, Gericke T, Mandel O, Bloch I (2006) New J Phys 8:152

Chapter 6
Motional Broadening in Ensembles with Heavy-Tail Detuning Distribution

In previous chapters we have shown that fluctuations in the resonance frequency of the two-level systems (TLS) cause narrowing of the spectrum. In other words, we have shown that fluctuations change the time-evolution of the phase difference between the two levels of the TLS from linear (ballistic) expansion to diffusion. In this chapter we address the question under what conditions the fluctuations can have the reverse effect to motional narrowing and lead to broadening of the spectral lines. An example for this effect was pointed out in Ref. [1]. We analyze the problem in a spectroscopic framework, and show that when the ensemble frequency distribution has heavy tails with a diverging mean, motional broadening emerges. In terms of quantum information, this manifests itself as a shortening of the coherence time as the the fluctuation rate increases. We derive a general equation for the linewidth of the spectrum, and demonstrate its validity through numerical simulations. Since in practice heavy tails of the frequency distribution can be sustained only up to some point, we study scenarios with cutoffs and show that motional broadening persists up to some fluctuation rate. Motional broadening is relevant to many fields in which heavy-tail distributions are encountered, including turbulence [2], diffusion [3] and laser-cooling [4].

We consider a similar model to the one introduced in Chap. 5 of an ensemble of two-level systems (TLS) which is described by the Hamiltonian given in Eq. (3.3). For this Hamiltonian the coherence can be written as

$$R(T) = |\langle e^{i\phi(T)}\rangle|, \tag{6.1}$$

where $\phi(T) = \int_0^T \delta(t)dt$ is the accumulated phase difference between the two internal states in the rotating frame. The spectrum, $S(\omega)$, is the absolute value squared of the Fourier transform of the coherence. The coherence start at 1 and decays to 0 as time advances. If the absolute value is omitted from the previous definition, oscillations may occur. Without fluctuations, the detuning of each TLS is constant in time, and the coherence is given by $R_0(T) = |\int_{-\infty}^{\infty} P_0(\delta)e^{i\delta T}d\delta|$.

Y. Sagi, *Collisional Narrowing and Dynamical Decoupling in a Dense Ensemble of Cold Atoms*, Springer Theses, DOI: 10.1007/978-3-642-29605-5_6, © Springer-Verlag Berlin Heidelberg 2012

6.1 The Case of Poisson Fluctuations

As explained already in Chap. 5, an explicit expression for $R(t)$ can be derived for fluctuations modeled by discrete spectral jumps and it is given in Eq. (5.1). This solution is valid for all $P_0(\delta)$, and in particular, to illustrate the transition from motional narrowing to broadening, we consider an ensemble with a Student's t-distribution:

$$P_0(\delta) = N(r, \delta_0) \left[1 + \frac{1}{r} \left(\frac{\delta}{\delta_0} \right)^2 \right]^{-\frac{1+r}{2}}, \tag{6.2}$$

with the normalization factor $N(r, \delta_0) = \Gamma\left(\frac{r+1}{2}\right) / \Gamma\left(\frac{r}{2}\right) \delta_0 \sqrt{r\pi}$, where $\Gamma(z)$ is the gamma function. For $r \to \infty$ the distribution is approaching a Gaussian with a standard deviation δ_0, and for $r = 1$ it is identical to the Cauchy distribution (Lorentzian). The first (second) moment of the distribution diverges for $r < 1$ ($r < 2$). Using Eq. (5.1), we calculate the spectrum and plot it in Fig. 6.1 for $r = 0.5$ and $r = 1.5$, with and without fluctuations. For $r = 1.5$ we observe that the spectrum becomes narrower in the presence of fluctuations, which demonstrate that motional narrowing persists even when the second moment diverges. On the other hand, for $r = 0.5$ the fluctuations broaden the spectrum. In Fig. 6.1 we plot as a function of Γ the normalized spectral width, which is defined to be the full width at half the maximum (FWHM) divided by the FWHM for $\Gamma = 0$. The figure clearly shows the narrowing (for $r = 1.5$) or broadening (for $r = 0.5$) effects as the fluctuation rate increases. Curiously, for the Cauchy distribution, corresponding to $r = 1$, there is no Γ dependency. This fact follows from Eq. (5.1), but it is also true regardless of the distribution of the collision times, as can be explained by the observations which follow. In Fig. 6.2 we plot the normalized FWHM as a function of r for various values of Γ. This figures clearly demonstrate that the transition point from motional narrowing to broadening is at $r = 1$.

6.2 Motional Broadening for Stable Distributions

It is instructive to consider the effect of fluctuations when the distribution of the detuning is given by one of the so-called 'stable laws' with a characteristic exponent $0 < \alpha \leq 2$ [5]. For each distribution in this class, the weighted sum of independent identically distributed variables produces a variable with a scaled version of the same distribution. More explicitly: for any two real numbers $q, s > 0$, and a pair of independent variables δ_1, δ_2 of such distribution, the weighted sum $(q\delta_1 + s\delta_2) / (q^\alpha + s^\alpha)^{1/\alpha}$ has the same distribution as δ_j. Some well-known examples of stable distributions are Gaussian ($\alpha = 2$), Cauchy ($\alpha = 1$) and Lévy ($\alpha = 1/2$) distributions. The coherence, as given in Eq. (6.1), is the absolute value of the characteristic function, $\varphi_\phi(t)$, of the phase distribution. Assuming P_0 is α-stable, the phase with discrete

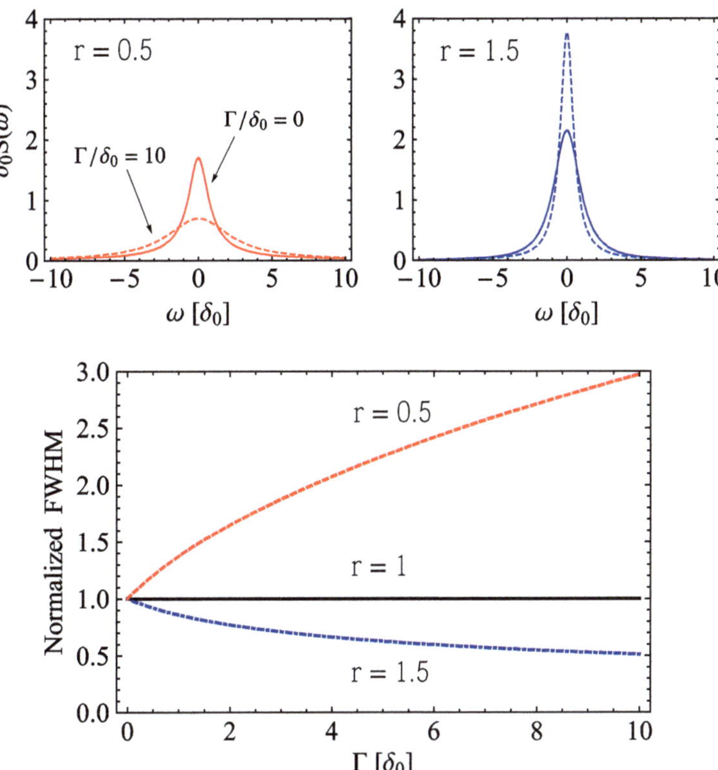

Fig. 6.1 The two upper graphs show the spectrum for a Student's t-distribution of the detunings [see Eq. (6.2)]. On the left the spectrum is plotted for $r = 0.5$ for which the first moment of the distribution diverges whereas on the right the spectrum is plotted for $r = 1.5$ for which only the second moment diverges but the first moment exists. For both spectrums the spectrum is plotted without fluctuations (*solid line*) and with fluctuations at $\Gamma = 10\delta_0$ (*dashed line*). In the lower graph we plot the full width at half the maximum (FWHM) normalized to the FWHM without fluctuations, as a function of Γ

fluctuations ('collisions') can be written as $\phi(T) = \sum_{j=1}^{n} \Delta t_j \delta_j \overset{D}{=} (\sum_{j=1}^{n} \Delta t_j^{\alpha})^{1/\alpha}\delta$, with Δt_j being the periods between collisions. For stable distributions the characteristic function satisfies: $|\varphi_\phi(t)| = e^{-c_\alpha |t|^\alpha}$, with some $c_\alpha > 0$ [5]. Thus, the coherence without collisions is given by $R_0(T) = e^{-c_\alpha T^\alpha}$ and with collisions it is given by

$$R(T) = e^{-c_\alpha (\sum_{j=1}^{n} \Delta t_j^\alpha)}. \tag{6.3}$$

For any series of collisions: $\sum_{j=1}^{n} \Delta t_j = T$, and hence:

$$\frac{\sum_{j=1}^{n} \Delta t_j^\alpha}{T^\alpha} = \sum_{j=1}^{n} \frac{\Delta t_j}{T}\left(\frac{\Delta t_j}{T}\right)^{\alpha-1} \begin{cases} \leq \tau_{max}^{\alpha-1} & \alpha \geq 1 \\ \geq \tau_{max}^{\alpha-1} & \alpha \leq 1 \end{cases} \tag{6.4}$$

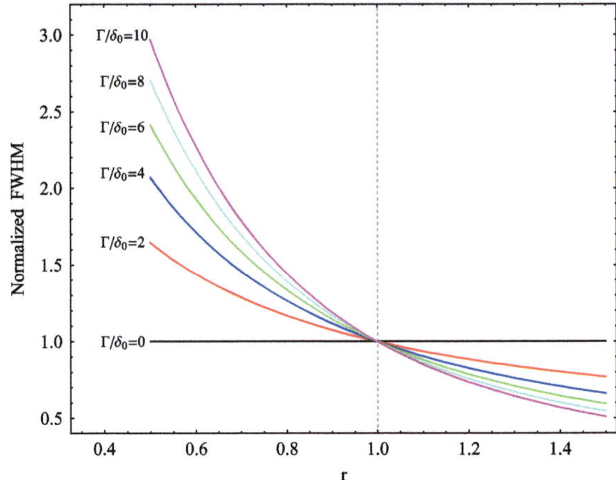

Fig. 6.2 The normalized spectral width as a function of r for different values of Γ. All lines cross at $r = 1$, above which there is motional narrowing and below motional broadening

with $\tau_{max} = \max\{t_j/T\}$. Combining Eqs. (6.3) and (6.4) and the fact that $\tau_{max} \leq 1$ we get

$$R(T) \begin{cases} > R_0(T), & \alpha > 1 \;\; \text{(motional narrowing)} \\ < R_0(T), & \alpha < 1 \;\; \text{(motional broadening)}. \end{cases} \tag{6.5}$$

with no other assumption regarding the distribution of the collision times.

6.3 Generalization

We now extend the above observations further to distributions which are by themselves not stable, but are in the domain of attraction of a stable distribution (or 'law') S_α. This notion means that a sum of variables drawn from the distribution, up to a normalizing factor, converges in distribution to S_α, as the number of summands increases. A distribution belongs to the domain of attraction of an α-stable law if its cumulative distribution function, $F(x)$, scales as $F(x) \sim |x|^\alpha h(|x|)$ as $x \to -\infty$, and $1 - F(x) \sim x^\alpha h(x)$ as $x \to \infty$, where $h(x)$ is a slowly varying function at infinity [6]. The domain of attraction of the Gaussian distribution contain all distributions with a finite variance.

If P_0 belongs to the domain of attraction of an α-stable distribution, then its characteristic function is of the form $|\varphi_\phi(t)| = e^{-c|t|^\alpha \tilde{h}(t)}$, with $\tilde{h}(t) = e^{o(t)}$ a slowly varying function as $t \to 0$ [6]. The coherence without collisions is given by $R_0(T) = \exp\left[-cT^\alpha \tilde{h}(T)\right]$, and with collisions it is given by $R(T) = \exp\left[-c\sum_{j=1}^n \Delta t_j^\alpha \tilde{h}(\Delta t_j)\right]$. Extending Eq. (6.4) one may see that if $\lim_{n\to\infty} \tau_{max} = 0$, at a fixed T, than for $\alpha > 1$: $\sum_{j=1}^n \Delta t_j^\alpha \tilde{h}(\Delta t_j) < C_1 T^\alpha \tau_{max}^{\eta_1}$, and for $\alpha <$

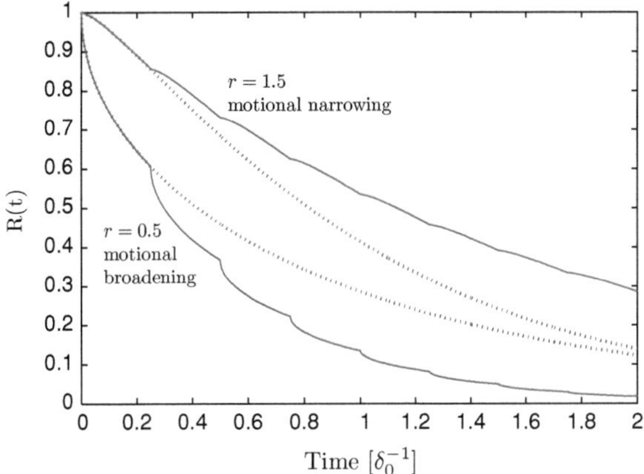

Fig. 6.3 The coherence without fluctuations (*dotted lines*) and with fluctuations separated by $\Delta t_j = 0.25 \cdot \delta_0^{-1}$ (*solid lines*). The detuning distribution is assumed to be the Student's t-distribution with $r = 0.5$ and 1.5. Since the coherence is given by $R(T) = R_0(\Delta t_1) \cdot R_0(\Delta t_2) \cdots R_0(\Delta t_n)$, a Zeno/anti-Zeno like behavior explains the transition from motional narrowing to broadening for a diverging first moment of P_0, at which point $\partial_T R_0|_{T=0}$ changes from 0 to $-\infty$

1: $\sum_{j=1}^{n} \Delta t_j^{\alpha} \tilde{h}(\Delta t_j) > C_2 T^{\alpha} \tau_{max}^{-\eta_2}$, with some $\eta_j, C_j > 0$. This extends the validity of Eq. (6.5) for the limit of many randomizing events (high collision rate) when the detuning have a distribution in the domain of attraction of an α-stable law.

The above can be summarized by saying that whether the coherence of a TLS ensemble decays faster or slower due to 'resetting' discrete fluctuations depends on the the tails of the detunings distribution. Motional broadening emerges for heavy-tailed distributions with $\alpha < 1$ which corresponds to a diverging first moment. This explains the results of Fig. 6.1 since the Student's t-distribution belongs to the domain of attraction of an α-stable distribution with $\alpha = r$. Furthermore, the criterion for motional broadening given in Eq. (6.5) coincides with the divergence of the detuning distribution's first moment.

To get an intuition for the above results we write the coherence at a time T, for a given series of randomization events, as

$$R(T) = \prod_{l=1}^{n} \int_{-\infty}^{\infty} d\delta_l P_0(\delta_l) e^{i \Delta t_l \delta_l} = R_0(\Delta t_1) \cdot R_0(\Delta t_2) \cdots R_0(\Delta t_n). \tag{6.6}$$

In addition, the derivative $\partial_T R_0|_{T=0^+}$ is 0 for $\alpha > 1$ and $-\infty$ for $\alpha < 1$. The combination of these two properties explains why depending on whether α is larger or smaller than 1, the coherence after a fluctuation lie above or below the original curve of $R_0(T)$, as depicted in Fig. 6.3 for the simple case of equal times between randomization events. In this respect, motional narrowing is analogous to the Zeno effect [7], and motional broadening to the anti-Zeno effect.

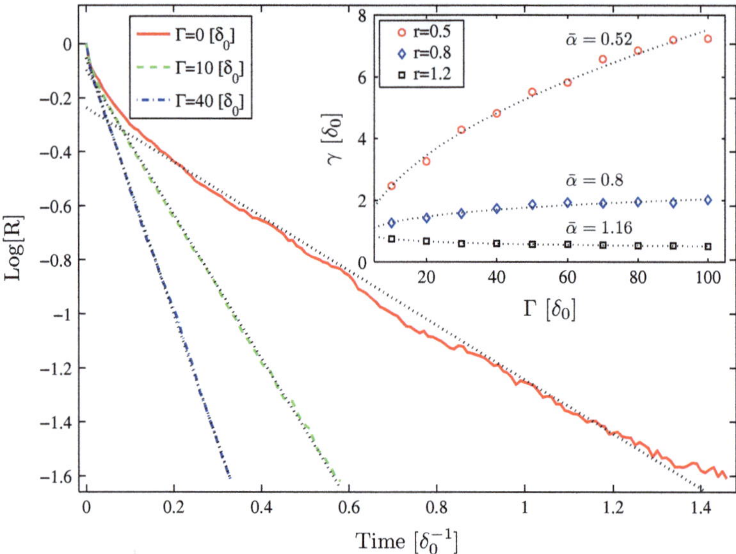

Fig. 6.4 Numerical simulation of the logarithm of the coherence using an ensemble of 10,000 particles following the Poisson discrete fluctuations model with three different rates Γ and detunings following a Student's t-distribution with $r = 0.5$. The *dotted* (*black*) *lines* are linear fits, validating the prediction of Eq. (6.7) that as Γ increases the decay becomes exponential. We extract the decay rate, γ, by fitting the simulated coherence for $T > 10\Gamma^{-1}$ to an exponentially decaying function $Ae^{-\gamma T}$. The inset shows γ as a function of Γ for different values of the distribution parameter r. The *dotted lines* are fits to the function $\gamma = a\Gamma^{1-\bar{\alpha}} + b$ (which has the functional form derived in Eq. (6.7). The extracted exponents agree well with the expected values $\bar{\alpha} = r$

We rewrite Eq. (6.3) using the typical time between collisions $\Delta t_j \sim \Gamma^{-1}$, and the inhomogeneous decay rate $\gamma_0 = c_\alpha^{1/\alpha}$, and obtain in the limit of many collisions $\Gamma T \gg 1$: $R(T) \approx e^{-\gamma_0^\alpha \Gamma^{1-\alpha} T}$. The asymptotic behavior of the coherence decays exponentially with time, and the decay rate is given by

$$\gamma = \gamma_0^\alpha \Gamma^{1-\alpha}. \tag{6.7}$$

This equation with $\alpha = 2$ reduces to the result obtained in Eq. (4.3). Since Eq. (6.3) is true only for a stable distribution, we test the validity of Eq. (6.7) for distributions in the domain of attraction of an α-stable distribution in numerical simulations. The results of these simulations for a Student's t-distribution are plotted in Fig. 6.4, and show that as the fluctuation rate increases the decay of coherence indeed becomes exponential. From the numerical curves we extract the decay rate and plot it in the inset of Fig. 6.4 for various values of the distribution parameter r and Γ. The functional form of the decay rates confirms the prediction of Eq. (6.7) with the correct α.

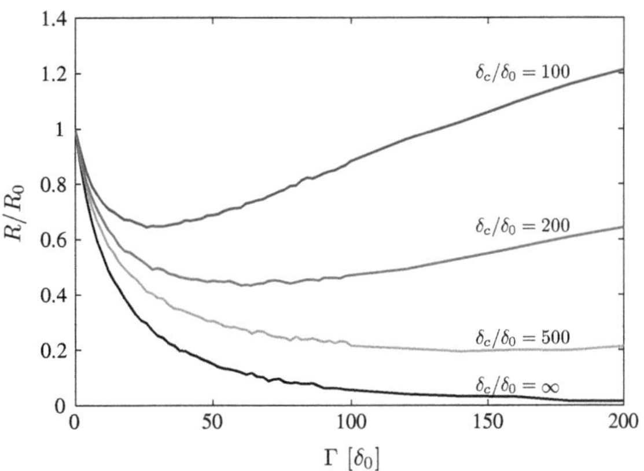

Fig. 6.5 Numerical simulation of the normalized coherence $R(T)/R_0(T)$ at time $T = 0.5\delta_0^{-1}$ versus the fluctuations rate Γ. The detuning distribution is taken to be the Student's t-distribution [see Eq. (6.2)] with $r = 0.5$, and the fluctuations are Poissonian. Each curve is calculated for P_0 truncated at a cutoff detuning δ_c

In real physical situations the detuning distribution can not have a diverging first moment, and the heavy tail scaling can be sustained up to some cutoff δ_c. For $P_0(\delta)$ with a characteristic exponent $\alpha < 1$, the order of magnitude of the sum $\sum_{j=1}^{n} \Delta t_j \delta_j$ is the same as of $\max\{t_j \delta_j\}$. The effect of the cutoff is therefore negligible as long as $\mathrm{Prob}(\max[\{\delta_j\}] > \delta_c) \ll 1$ (we assume $\langle \Delta t_j \rangle = \Gamma^{-1} < \infty$). This probability depends on the number of collisions, which is roughly given by ΓT. An estimate of this probability yields that the effect of the cutoff is insignificant for $\Gamma T \ll (\delta_c/\delta_0)^{\alpha}$, where δ_0 is the typical scale of the detuning distribution [see Eq. (6.2)]. This means that for a given observation time T, motional broadening persists up to fluctuation rate on the order of $(\delta_c/\delta_0)^{\alpha} T^{-1}$. This qualitative picture is demonstrated in numerical simulations plotted in Fig. 6.5. Motional broadening prevails for small Γ, later changing to motional narrowing once the cutoff is sampled and discovered.

The TLS ensemble coherence problem can be mapped to that of particles performing diffusion in free-space, where the detuning and accumulated phase are mapped to velocity and position, respectively. In this analogy, the diffusion problem assumes an ensemble of particles with a steady-state distribution of velocities, starting all from the same point in space. In the absence of collisions, the particles are ballistically expanding and the width of their position distribution grows linearly with time. With collisions, our criterion yields that the width of the particles' spatial distribution grows faster than ballistic (super-ballistic) for heavy-tailed velocity distribution. Super-ballistic diffusion is known to exist in turbulent flow [8]. In the context of spatial diffusion, a particularly interesting implementation of the model considered in this paper can be achieved. It was shown that the steady-state velocity distribution of atoms in a polarization lattice follows a power law with an exponent that depends on the lattice depth [9]. As a result, in such a system the diffusion becomes anomalous [10]. Going to a low enough lattice depth, it should be possible to observe motional broadening, namely diffusion whose scaling with time is faster than ballistic.

References

1. Burnstein A (1981) Chem Phys Lett 83:335
2. Falkovich G, Gawedzki K, Vergassola M (2001) Rev Mod Phys 73:913
3. Bouchaud J-P, Georges A (1990) Phys Rep 195:127
4. Bardou F, Bouchaud J, Aspect A, Cohen-Tannoudji C (2002) Lévy statistics and laser cooling. Cambridge University Press, Cambridge
5. Feller W (1968) An introduction to probability theory and its applications, vol II, 3rd edn. Wiley, New York
6. Ibragimov IA, Linnik YV (1971) Independent and stationary sequences of random variables. Wolters-Noordhoff, Groningen
7. Milburn GJJ (1988) Opt Soc Am B 5:1317
8. Shlesinger MF, West BJ, Klafter J (1987) Phys Rev Lett 58:1100
9. Castin Y, Dalibard J, Cohen-Tannoudji C (1991) The limits of Sisyphus cooling. In: Moi L, Gozzini S, Gabbanini C, Arimondo E, Strumia F (eds.) Light induced kinetic effects on atoms, ions, and molecules. ETS Editrice, Pisa
10. Marksteiner S, Ellinger K, Zoller P (1996) Phys Rev A 53:3409

Chapter 7
Suppression of Collisional Decoherence by Dynamical Decoupling

7.1 Introduction

As was already mentioned, cold atomic ensembles can be used as an interface between matter and photonic qubits in quantum networks, and in recent years vast experimental advances in this direction have been reported [1–6]. The effect of collisional fluctuations was analyzed in the previous chapters and the decoherence it induces is well understood. Though fluctuations at low frequencies can be overcome by a single population inverting pulse—the celebrated coherence echo technique [7, 8], as the collision rate increases this is no longer possible due to higher frequency components. Dynamical decoupling (DD) theories generalize this technique to multi-pulse sequences by harnessing symmetry properties of the coupling Hamiltonian [9–13]. Though DD was demonstrated in several experiments [14–19], its use with atomic ensembles remains unexplored to date. In this chapter we study experimentally DD in a dense cold atomic ensemble and report on a substantial suppression of collisional decoherence.

We again consider atoms with internal states $|1\rangle$ and $|2\rangle$, trapped in a conservative optical potential which are described by the effective Hamiltonian given in Eq. (3.3). We have shown that the state of each atom is given by $|\psi(t)\rangle = 2^{-1/2}(|1\rangle + e^{-i\phi(t)} |2\rangle)$, where the phase difference is given by $\phi(t) = \int_0^\infty \delta(t)dt$. A schematic plot of three realizations of $\phi(t)$ is given in Fig. 7.1 (top), and it can be seen that the phase difference is accumulated in a constant rate between collisions [20]. The effect of a population inverting pulse (π-pulse) is to change the sign of δ, and a train of such pulses lead to a much narrower phase distribution and slower decoherence, as depicted in Fig. 7.1 (bottom).

Y. Sagi, *Collisional Narrowing and Dynamical Decoupling in a Dense Ensemble of Cold Atoms*, Springer Theses, DOI: 10.1007/978-3-642-29605-5_7,

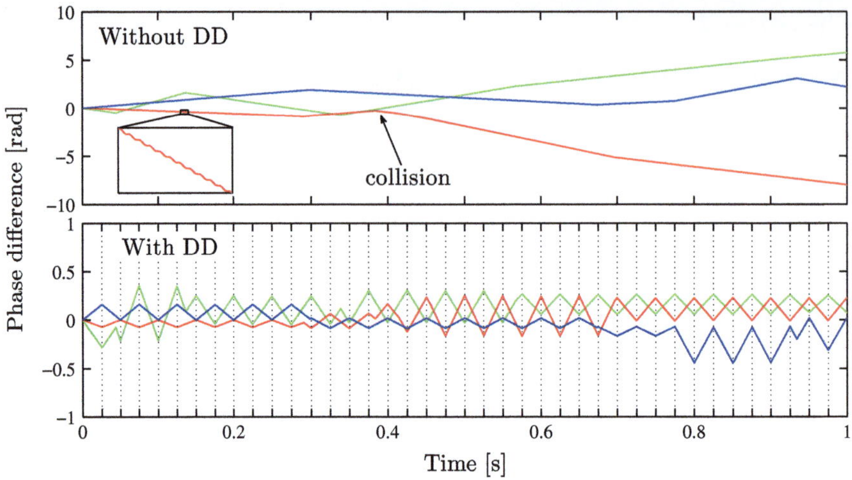

Fig. 7.1 Schematic drawing of the evolution of the relative phase between the two internal states. Without dynamical decoupling (*top*), the average potential energy of the atom is changed after a collision and therefore also its average rate of phase accumulation, δ. With dynamical decoupling pulses (*dotted lines*) the spread of the phases is much smaller (note the different graph scales). The inset shows small oscillations due to the fast periodic atomic motion in the trap. Since the oscillation period is shorter than all relevant timescales in our experiment, we consider only $\delta(t)$ averaged over an oscillation period [20]

7.2 Quantum Process Tomography of Dynamical Decoupling

The conditions in the experiments presented here are 275,000 atoms at a temperature of $1.7\,\mu$K, phase space density of 0.04 and an average collision rate of $100\,\mathrm{s}^{-1}$. The typical inhomogeneous decay time as measured in a Ramsey-like experiment is $\sim$150 ms. The peak optical depth for a non-polarized resonant light is $\sim$230.

We employ a Carr-Purcell-Meiboom-Gill (CPMG) decoupling scheme [21] and show in what follows that for collisional detuning fluctuations it is virtually optimal. The pulse sequence is composed of n π-pulses at times $t_k = \frac{2k-1}{2n}t$ where $k = 1 \ldots n$ (see Fig. 7.2), and we characterize it by the effective frequency $f_{DD} = \frac{n}{2t}$. We study the effect of the DD scheme by performing a quantum process tomography (QPT) [22]. QPT enables us to reconstruct the χ-matrix which gives a convenient way to calculate the density matrix after the process, ρ_{out}, in terms of the initial density matrix, ρ_{in}, by $\rho_{out} = \mathcal{E}[\rho_{in}] = \sum_{k,l} \hat{E}_k \rho_{in} \hat{E}_l^\dagger \chi_{kl}$, where $\hat{\mathbf{E}} = (\hat{I}, \hat{X}, -i\hat{Y}, \hat{Z})$ with $(\hat{I}, \hat{X}, \hat{Y}, \hat{Z})$ being the Pauli matrices. For a single qubit this is conveniently done by a measurement of $tr[\hat{X}\rho]$, $tr[\hat{Y}\rho]$ and $tr[\hat{Z}\rho]$. We measure the population at state $|2\rangle$ with and without another π pulse, and the difference of the two values normalized to the initial population at state $|1\rangle$ gives $\frac{tr[\hat{Z}\rho]+1}{2}$. We obtain the other two projections by applying a $\pi/2$ pulse in the suitable axis and then measure $tr[\hat{Z}\rho]$.

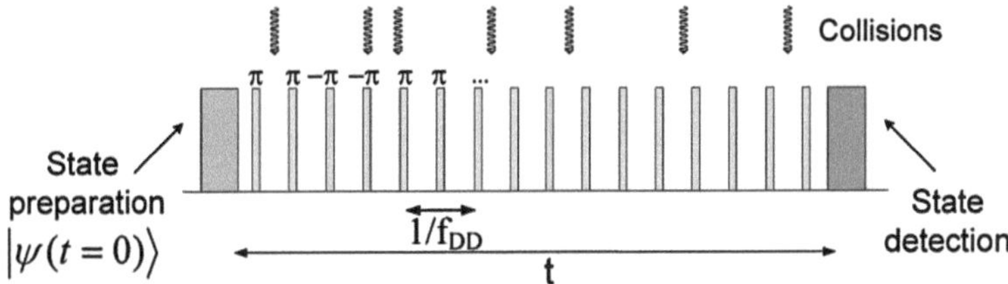

Fig. 7.2 The experimental pulse sequence starts with a state preparation, followed by a train of π-pulses with alternated phases to minimize the accumulation of errors due to pulse width and frequency inaccuracies, and a final state detection. The duration of each π pulse is $\sim$0.5 ms and its average fidelity is $\sim$0.995

The results of a QPT of a DD sequence with $f_{DD} = 35$ Hz are depicted in Fig. 7.3. There are two distinctive decay timescales for the equatorial plane and the z-axis, which correspond to phase damping noise processes and depolarizing noise processes (T_1), respectively. The former originates from collisional fluctuations in δ and it is the dominant noise process which determines the ensemble coherence time, τ_c, which is quantified by $R(t)$. Depolarization process is induced by inelastic collisions, and its typical timescale is measured to be $T_1 = 6$ s [23]. The worst case fidelity of the ensemble as a quantum memory, defined as $\mathcal{F} = \min_{|\psi\rangle}\langle\psi|\mathcal{E}\left[|\psi\rangle\langle\psi|\right]|\psi\rangle$, is calculated from the measured χ-matrix to be $\mathcal{F} = 0.83$, 0.74 and 0.64 for 1, 2 and 3 s, respectively, which corresponds to an exponential decay timescale of $\tau_c = 2.4$ sec. The contraction of the Bloch sphere is symmetric in the equatorial plane, which indicates that the decoupling scheme is insensitive to the stored superposition. We demonstrate this point with a direct measurement in which we start with two orthogonal initial states in the equatorial plain and scan the phase of a final $\pi/2$ pulse added to the sequence and measure the population at $|2\rangle$ normalized to its initial value. The results depicted in the inset of Fig. 7.4 exhibit the same contrast and preserve the $\pi/2$ phase shift between the two initial states.

7.3 Optimal Decoupling Sequence for Collisional Bath

The decay of the coherence with a DD pulse sequence and assuming a Gaussian phase distribution is given in a system-reservoir framework by Eq. (3.5) with a filter function which encapsulates the information on the DD pulse sequence and is given by $F(\omega t) = \frac{1}{2}|\sum_{k=0}^{n}(-1)^k(e^{i\omega t_{k+1}} - e^{i\omega t_k})|^2$ with $t_0 = 0$ and $t_{n+1} = t$. As already explained, the power spectrum is given by $S_\delta(\omega) = \frac{2\Gamma\sigma_\delta^2}{\Gamma^2+\omega^2}$, where Γ^{-1} is the velocity correlation time. By solving numerically Eq. (3.5) and leaving the $\{t_i\}_{i=1}^{n}$ as free parameters, we find that the optimal decoupling sequence for a Lorentzian power spectrum is given by $t_i = \frac{\eta+i-1}{n-1+2\eta}t$, where $i = 1...n$ and $0.5 \leq \eta \leq 1$ is a numerical factor which depends on n and t. For $\frac{\Gamma t}{n} \ll 1$ we find $\eta \approx 0.5$, for which we retrieve

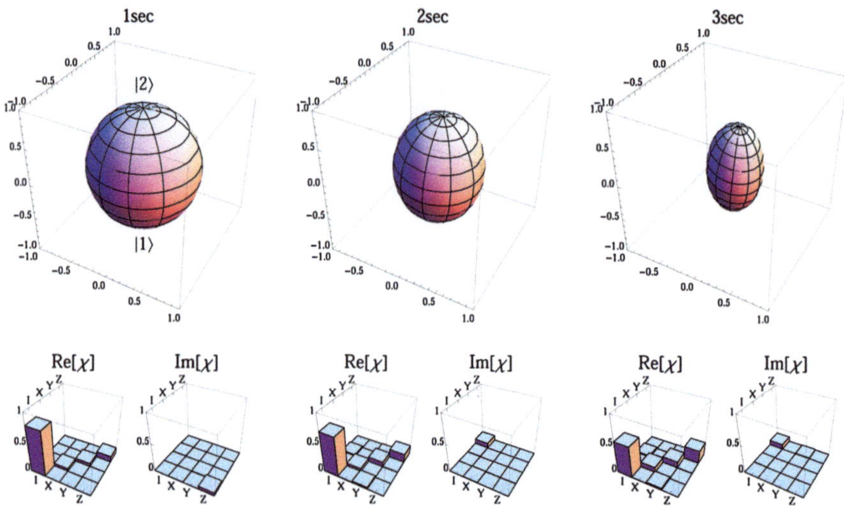

Fig. 7.3 Quantum process tomography of dynamical decoupling with $f_{DD} = 35\,\text{Hz}$. Any single-qubit density matrix ρ can be mapped to a point in space $\bar{r}(\rho) = (tr(\hat{X}\rho), tr(\hat{Y}\rho), tr(\hat{Z}\rho))$, with $(\hat{X}, \hat{Y}, \hat{Z})$ being the Pauli matrices. The colors and lines are chosen for the initial states, ρ_{in}, which lie on a sphere with a radius 1. For each of these states we calculate the process outcome, $\rho_{out} = \mathcal{E}[\rho_{in}]$, and plot it with its initial color at $\bar{r}(\rho_{out})$. The contraction of the Bloch sphere is more pronounced on the equatorial plane which shows that the main noise process is phase damping, as descried by the Hamiltonian of Eq. (3.3). There is also a rotation of the sphere around the $|1\rangle$-$|2\rangle$ axis at a rate of $\sim 9\,^\circ\,\text{s}^{-1}$ due to small inaccuracies in the control field

the CPMG pulse sequence. Furthermore, even when $\frac{\Gamma t}{n} \approx 1$ the coherence time with the CPMG pulse sequence differs by less than 1% from the optimal value. We have tested theoretically and experimentally other DD schemes, and in particular the one suggested in Ref. [12], and verified that they are indeed inferior to the CPMG sequence in our Lorentzian fluctuations power spectrum (for more details see Appendix D).

7.4 Coherence Time Measurements with Dynamical Decoupling

We measure $R(t)$ directly by preparing the atoms in the superposition $|\psi\rangle = \frac{1}{\sqrt{2}}(|1\rangle + |2\rangle)$, employ the DD pulse sequence and finally measure the length of the Bloch vector with quantum state tomography. Though $R(t)$ does not have to follow, a priori, some well defined function, the experimental results depicted in Fig. 7.4 show that the data is well fitted by an exponentially decaying function e^{-t/τ_c}, from which we extract the coherence time τ_c. The exponential decay is expected in the Markovian limit, where the decay timescale is much larger than the fluctuations correlation time [13].

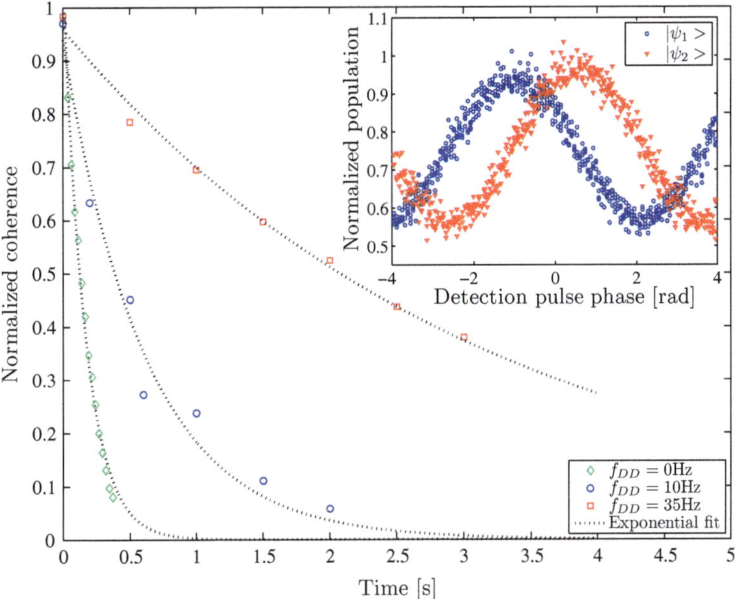

Fig. 7.4 The ensemble coherence versus time without DD ($f_{DD} = 0$ Hz) and with two representing DD pulse rates, normalized to the initial coherence. The inset shows a storage of two orthogonal initial states in the equatorial plane: $|\psi_1\rangle = \frac{1}{\sqrt{2}}(|1\rangle + |2\rangle)$ and $|\psi_2\rangle = \frac{1}{\sqrt{2}}(|1\rangle + e^{i\pi/2}|2\rangle)$. We add to the decoupling scheme another pulse independent of the initial state with a phase and duration chosen to correct for the small rotation of the Bloch sphere as was measured in the QPT. We measure the population at $|2\rangle$ after 3 s, normalized to the initial population, versus the phase of a $\pi/2$ detection pulse. The fringe contrast is not centered to 0.5 due to inelastic m-changing transitions

A measurement of the dependance of the coherence time on the DD pulse rate is shown in Fig. 7.5. We observe a quadratic increase of the coherence time versus f_{DD} up to 35 Hz, for which there is a 20-fold improvement to more than 3 s.

In order to explain these results we present a qualitative model for the coherence time. Without collisions the inhomogeneous dephasing rate is proportional to σ_δ. For simplicity we assume that if a collision did not occur between two consecutive π-pulses the inhomogeneous broadening is averaged out. If a collision occurred, however, the width of the ensemble phase distribution increases by $\sim f_{DD}^{-1}\sigma_\delta$. The number of collisions up to a time t is $\Gamma_{col}t$, and since we add random variables (i.e. the accumulated phase), the width of the phase distribution increases as a square root of time: $\Delta\Phi(t) \sim f_{DD}^{-1}\sigma_\delta\sqrt{\Gamma_{col}t}$. For cold collisions in 3D harmonic trap, the relation between the collision rate and the relaxation rate was shown to be $\Gamma_{col} = 2.7 \cdot \Gamma$ [24]. The coherence time, τ_c, is the time for which the width of phase distribution is on the order of 1, yielding

$$\tau_c \sim f_{DD}^2\sigma_\delta^{-2}\Gamma^{-1}, \tag{7.1}$$

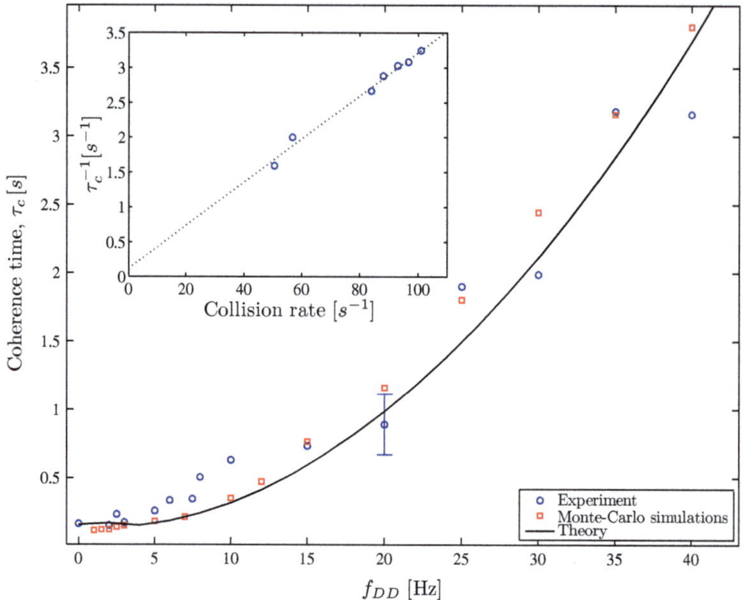

Fig. 7.5 The coherence time versus the dynamical decoupling pulse rate f_{DD}. The experimental data (*blue circles*) agrees well both with the theoretical prediction of Eq. (3.5) taking a Lorentzian power spectrum (*solid line*) with $\sigma_\delta = 23.8\,\mathrm{s}^{-1}$ and $\Gamma = 37.5\,\mathrm{s}^{-1}$ which were extracted from measured quantities and with no fit parameters, and with molecular dynamics Monte-Carlo simulations done with 1000 atoms in a 3D harmonic trap with an inhomogeneous decay time similar to the experiment (*red squares*). The errorbar is a estimated from the fits shown in Fig. 7.4. The inset shows a measurement of the dephasing rate for different collision rates for $f_{DD} = 8\,\mathrm{Hz}$, demonstrating the linear dependence of Eq. (7.1). The dotted line is a linear fit of the data

with a parabolic dependence on f_{DD}. This result can be also obtained from Eq. (3.5) by approximating $\frac{F(\omega t)}{\pi(\omega t)^2} \approx \delta_{Dirac}(\omega t - 2\pi f_{DD}t)$ and using the Lorentzian power spectrum.

Exact calculations of τ_c using Eq. (3.5) without fitting parameters are presented in Fig. 7.5 in good agreement with the experimental data. The calculations are done with a Lorentzian power spectrum where the parameters Γ and σ_δ are extracted from measured quantities. Γ is extracted from the collision rate which is calculated using the measured temperature, number of atoms and trap oscillation frequencies. As mentioned before, the parameter σ_δ is measured in a Ramsey experiment at very low densities, where the collisions can be disregarded and σ_δ can be extracted from the measured dephasing rate. We also perform Monte-Carlo simulations, and its results are also depicted in Fig. 7.5, and agree well with both theory and experiments. We conclude that the effect of collisions can be indeed formulated as an effective single spin Hamiltonian coupled to a reservoir. Moreover, although the detunings of atoms trapped in a 3D harmonic trap are not normally distributed, the distribution of their accumulated phase can be well approximated by a Gaussian owing to the central limit theorem and the large number of collisions involved.

Another prediction of Eq. (7.1) is the linear dependence of the coherence time on Γ^{-1}. In the experiment we change Γ by reducing the density and collision rate while keeping the temperature, and therefore σ_δ, unchanged [20]. This is accomplished by reducing the intensity of the cooling lasers in the trap loading stage. In the inset of Fig. 7.5 we plot τ_c^{-1} versus the average collision rate for a pulse rate of $f_{DD} = 8\,\mathrm{Hz}$. As expected, the coherence time is inversely proportional to the collision rate.

References

1. Kuzmich A, Bowen WP, Boozer AD, Boca A, Chou CW, Duan LM, Kimble HJ (2003) Nature 423:731
2. Chou CW, de Riedmatten H, Felinto D, Polyakov SV, van Enk SJ, Kimble HJ (2005) Nature 438:828
3. Yuan Z, Chen Y, Zhao B, Chen S, Schmiedmayer J, Pan J (2008) Nature 454:1098
4. Zhao B, Chen YA, Bao XH, Strassel T, Chuu CS, Jin XM, Schmiedmayer J, Yuan ZS, Chen S, Pan JW (2009) Nat Phys 5:95
5. Schnorrberger U, Thompson JD, Trotzky S, Pugatch R, Davidson N, Kuhr S, Bloch I (2009) Phys Rev Lett 103:033003
6. Zhang R, Garner SR, Hau LV (2009) Phys Rev Lett 103:233602
7. Hahn EL (1950) Phys Rev 80:580
8. Andersen MF, Kaplan A, Davidson N (2003) Phys Rev Lett 90:023001
9. Viola L, Lloyd S (1998) Phys Rev A 58:2733
10. Viola L, Knill E, Lloyd S (1999) Phys Rev Lett 82:2417
11. Search C, Berman PR (2000) Phys Rev Lett 85:2272
12. Uhrig GS (2007) Phys Rev Lett 98:100504
13. Cywinski L, Lutchyn RM, Nave CP, Sarma SD (2008) Phys Rev B 77:174509
14. Fortunato EM, Viola L, Hodges J, Teklemariam G, Cory DG (2002) New J Phys 4:5
15. Fraval E, Sellars MJ, Longdell JJ (2005) Phys Rev Lett 95:030506
16. Morton JJL, Tyryshkin AM, Ardavan A, Benjamin SC, Porfyrakis K, Lyon SA, Briggs GAD (2006) Nat Phys 2:40
17. Damodarakurup S, Lucamarini M, Di Giuseppe G, Vitali D, Tombesi P (2009) Phys Rev Lett 103:040502
18. Biercuk MJ, Uys H, VanDevender AP, Shiga N, Itano WM, Bollinger JJ (2009) Nature 458:996
19. Du J, Rong X, Zhao N, Wang Y, Yang J, Liu RB (2009) Nature 461:1265
20. Kuhr S, Alt W, Schrader D, Dotsenko I, Miroshnychenko Y, Rauschenbeutel A, Meschede D (2005) Phys Rev A 72:023406
21. Vandersypen LMK, Chuang IL (2005) Rev Mod Phys 76:1037
22. Nielsen MA, Chuang IL (2000) Quantum computation and quantum information. Cambridge University Press, Cambridge
23. Widera A, Gerbier F, Folling S, Gericke T, Mandel O, Bloch I (2006) New J Phys 8:152
24. Monroe CR, Cornell EA, Sackett CA, Myatt CJ, Wieman CE (1993) Phys Rev Lett 70:414

Chapter 8
Summary and Outlook

8.1 Summary

In this thesis I have studied an ensemble of trapped ultra-cold atoms. Atomic ensembles can be used as quantum memories in applications such as quantum computation and long-distance quantum networks. In a typical experiment, information contained in an electromagnetic field is mapped into the coherence between the two internal levels of the atoms. A high optical depth of the ensemble enhances its coupling to the light field and improves the overall efficiency of the storage and retrieval process. It is therefore plausible to use optically thick and consequently dense atomic media. Starting with a traveling pulse of light, the conversion process leaves the atomic coherence at different positions with different phases. When the information is retrieved, this spatially dependent phase is crucial for a full and correct reconstruction of the photonic pulse. In this thesis, however, we consider only a small volume where the stored coherence can be considered to have a uniform phase.

The atomic ensemble is an interacting many body system. We restrict our study to low phase space densities where the ensemble can be treated classically. Also, since the interactions are weak we adopt a mean field approach and use an effective single spin Hamiltonian. Our system is therefore described by a reduced density matrix, and the coherence is given by its off diagonal elements. The situation we envision is that the ensemble is prepared in an initial coherent superposition, and the reduced density matrix describes a pure state. As time progresses, however, the coherence decays. This happens even for low density ensembles due to the inhomogeneous broadening in the two-level energy difference. If we denote this bare dephasing rate by γ and the elastic collision rate by Γ, we differentiate between two regimes; the first regime is when $\Gamma \ll \gamma$, and collisions play no role in the coherence dynamics. In this case, decoherence can be largely eliminated by the well-known echo technique. The other regime will be the starting point of this work, where Γ is at least comparable if not larger than the bare dephasing rate. The questions I have addressed in this context are: what are the functional forms and the asymptotic decay time of the coherence? How does this depend on the specific collision model? can we have both narrowing

Y. Sagi, *Collisional Narrowing and Dynamical Decoupling in a Dense Ensemble of Cold Atoms*, Springer Theses, DOI: 10.1007/978-3-642-29605-5_8,
© Springer-Verlag Berlin Heidelberg 2012

and broadening effects due to the collisions? And finally, can the coherence time be extended by the application of a train of spin flipping pulses known as dynamical decoupling?

To answer these questions I have built a new experimental setup, in which lasers are used to gather and cool ^{87}Rb atoms to micro Kelvin temperatures, and thereupon trap them at high densities in an optical dipole potential. The thermodynamic conditions of the ensemble are characterized by taking absorption images of its spatial density distribution in situ and after a time-of-flight. Using microwave and radio-frequency fields, one can manipulate the internal degrees of freedom of the atoms, and subsequently measure them using state-selective fluorescence techniques. Employing these in time-domain Ramsey experiments, one can measure the ensemble spectrum of a specific "clock" transition between two meta-stable hyperfine states.

I have first studied the asymptotic coherence time of the ensemble. I have shown that velocity-changing elastic collisions prolong the coherence time, or putting it differently, narrow the linewidth. This phenomenon is very similar to the motional narrowing effect first observed in NMR. I have showed theoretically and confirmed experimentally that the emerging longer coherence time universally depends only on the atomic phase space density. These findings are further supported by classical molecular dynamics Monte-Carlo simulations.

Next, I have studied how this effect depends on the physical model of the fluctuations. To this end I have considered a fluctuation model in which the ensemble is coupled to an environment which induces random jumps separated by times which have a Poisson distribution. In this model a closed-form formula for the spectrum in terms of the inhomogeneous spectrum and the Poisson rate constant exists. As a case study we looked at a Gaussian detuning distribution and found that the spectrum depends on the softness of the collision, where in "softness" I mean how many collisions are needed to randomize the detuning (hard collision corresponds to only one). I have also demonstrated experimentally that the discrete spectral jumps model correctly accounts for the spectrum arising from cold elastic collisions in an optically trapped ensemble, without fitting parameters.

Another benefit of the spectral jumps model is that the solution for the spectrum holds for any frequency distribution. In particular, I have found that for fat-tail distributions the effect is reversed, namely instead of narrowing of the spectrum fluctuations lead to broadening. Using ideas from the mathematical theory of sums of identical and independent variables, it can be proved that this behavior arises for frequency distributions which belong to the domain of attraction of a Lévy stable law with a characteristic exponent smaller than 1. This understanding can be extended to other related phenomena such as anomalous diffusion and the Zeno effect.

Finally, I have addressed the question of whether the coherence time of a collisional-narrowed atomic ensemble can be extended by applying external control fields. For this purpose I have used ideas from dynamical decoupling theories, which generalize the Hahn echo technique to multi-pulse sequences and enable the suppression of noise at a higher frequency. In the experiment I have demonstrated a 20-fold increase of the coherence time when a dynamical decoupling sequence with more than 200 pi-pulses was applied. A measurement of the coherence time as a function of pulses frequency agrees well with theory, assuming a Lorentzian spectral function for the fluctuations. This spectral function is expected for a Poisson process, and the theory only uses two parameters which are measured in the experiments. Using quantum process tomography it is shown that a dense ensemble with an optical depth of 230 can be used as an atomic memory with coherence times exceeding 3 seconds. In other words, the suppression demonstrated in the experiment does not depend on the specific initial coherence in the ensemble—a necessary requirement for memory.

8.2 Outlook

The discrete fluctuation model revealed many important features of the collisional narrowed or broadened spectrum, but it seems that the assumption of a complete loss of correlation after a collision is not really required. An important extension of discussions in this thesis is towards models which contain correlations. It is, in many ways, an extension of the theory towards a soft collision model. I think that the first step towards this goal would be to simply consider correlations in time which decay exponentially.

As I have explained, the shape of the spectrum was already demonstrated to depend on the softness of the collision process. This was studied in detail for the case of a Gaussian detuning distribution. This case, which may be the simplest to deal with theoretically, is also interesting from an experimental point of view. In hot atomic ensembles the dominant cause of inhomogeneous broadening is the Doppler effect, and the bare spectrum is very well approximated by a Gaussian (as long as the Doppler width is much higher than the natural width). A measurement of a Dicke narrowed spectrum will probe the effect of the softness of the collisions. The experiment we envision is a vapor cell in which one can choose the buffer gas species and control its pressure. By changing the mass ratio between the active atom and the buffer gas, the relative velocity change of the active atom after a single collision can be altered. The measurement should be done in the regime where the inhomogeneous spectrum is narrowed by approximately a factor of two. In this regime the difference in the spectrums of "soft" and "hard" collision model is maximized. To achieve this one can measure the EIT spectrum in a geometrical configuration where there is a small angle between the probe and pump beams. Alternatively, the pressure of the buffer gas can be reduced to decrease the Dicke factor.

Observing motional broadening in an experiment is a major challenge for the future. The main problem is how to create a heavy-tail distribution of detunings or velocities. If the ensemble is in thermal equilibrium, the energy distribution decays exponentially at large energies because of the Boltzmann factor. This prohibits any heavy-tail distribution for the velocities or detunings (since both are correlated to the energy of the atom). A necessary step, therefore, for achieving such distributions is to be in equilibrium with some external reservoir which is not by itself in thermal equilibrium. The experiment proposed in chap. 6 uses exactly this notion; atoms are left to equilibrate with a polarization lattice, similar to the one used in Sisyphus cooling. The steady state velocity distribution is a power law, with a power that depends on the depth of the lattice. Motional broadening will than manifest itself as a diffusion process at which the width of the atomic cloud grows faster than in the ballistic case. By turning off the lattice and leaving the atoms to freely expand it is possible to measure the velocity distribution directly. The advantage of such an experiment is that it enables one to measure independently super-ballistic spatial diffusion and heavy-tail momentum distributions.

Another promising future direction is to apply better and more sophisticated dynamical decoupling schemes and improve further the coherence time. One way to do it is to use pulses which are not exactly spin flipping. For a Lorentzian spectral function as we have in our system, we have found out recently in work done in collaboration with the group of Prof. Gershon Kurizki of the Weizmann institute that a train of fractional $\sim 0.8\pi$ pulses achieves better results. Experimentally, this experiment is more challenging owing to two reasons; first, the improvement over a sequence of π-pulses is relatively small (on the order of 5 %). Second, there is not a straight forward way to a correct for inaccuracies in the pulses such as the phase alternation technique employed in previous experiments. Ultimately, one would like to measure directly the spectral function of the bath and fit to it the optimal dynamical decoupling sequence. Two features which have been neglected in this thesis and are expected to appear in such a measurement are the effect of the rapid oscillatory motion of the atoms, and spatial correlations arising due to the non homogeneous density profile in the trap and the collision rate (on average there are more collisions in denser places).

The use of dynamical decoupling should not be limited to thermal ensembles or coherence which was created by microwave fields. In particular, the use of dynamical decoupling to increase the storage time of a pulse of light in a cold ensemble was never demonstrated. A question which arises in this context is whether we can choose a dynamical decoupling sequence better suited to suppress the decoherence of a stored light, since we know that to first order the atoms populate only a single state. Of course the atomic motion in our experiment will wash out spatial information, and this can be prevented by either working in an optical lattice, or by treating the ensemble as a single spatial qubit (for experiments with co-propagating pump and probe beams). It would also be interesting to use dynamical decoupling with a Bose-Einstein condensate, where the challenge is to effectively decouple differential mean-fields effects. These effects due to inter-particles interactions are expected to be dominant for a condensate.

Appendix A
Collisional Narrowing Data Analysis

In the general case where the detuning distribution is not a Gaussian it can be shown that the Ramsey decay envelope is given by [1]:

$$\tilde{R}(s) = \frac{\tilde{R}_0(s + \Gamma_{col})}{1 - \Gamma_{col}\tilde{R}_0(s + \Gamma_{col})}, \tag{A.1}$$

where $\tilde{R}(s)$ and $\tilde{R}_0(s)$ are the laplace transforms of the Ramsey with and without collisions, respectively. As seen in the previous chapter, for 3D harmonic trap the Ramsey without collisions can be calculated analytically, and using Eq. (A.1) we get the laplace of the Ramsey with collisions. Since analytic solution to the inverse laplace does not exist in this case, we use instead a Gumbel function that approximates the numerical solutions well (see Fig. A.1). We therefore fit the extracted Ramsey envelopes to the function

$$f(t, a, b, \tau_1, \tau_2) = \sqrt{a^2 R^2_{Gumbel}(t) + b^2},$$

where $R_{Gumbel}(t)$ is defined in Eq. (7) in the paper, a and b are global amplitude and noise bias, $\Gamma_{col} = \tau_2/(1.69\tau_1)^2$ in accordance with Eq. (5) for 3D harmonic trap and τ_1 is calculated from the measured temperature (see Appendix). Note that the only free parameters are a, b and τ_2. In order to eliminate systematic errors we refine our analysis in the following way: taking some Γ_{col} we compute numerically the correct $R(t)$ using Eq. (A.1). We then fit it with f and compare the value found for τ_2 to the correct value calculated from Eq. (5), and define a correction factor which is mostly in the range of $0.8 - 1.2$. We repeat this and construct a table of the correction factor as a function of $\Gamma_{col}\tau_1$ (see inset of Fig. A.1), and use this table to correct the experimentally fitted τ_2 values. It is important to stress, however, that this procedure does not change the slopes by more than 20 %.

Y. Sagi, *Collisional Narrowing and Dynamical Decoupling in a Dense Ensemble of Cold Atoms*, Springer Theses, DOI: 10.1007/978-3-642-29605-5, © Springer-Verlag Berlin Heidelberg 2012

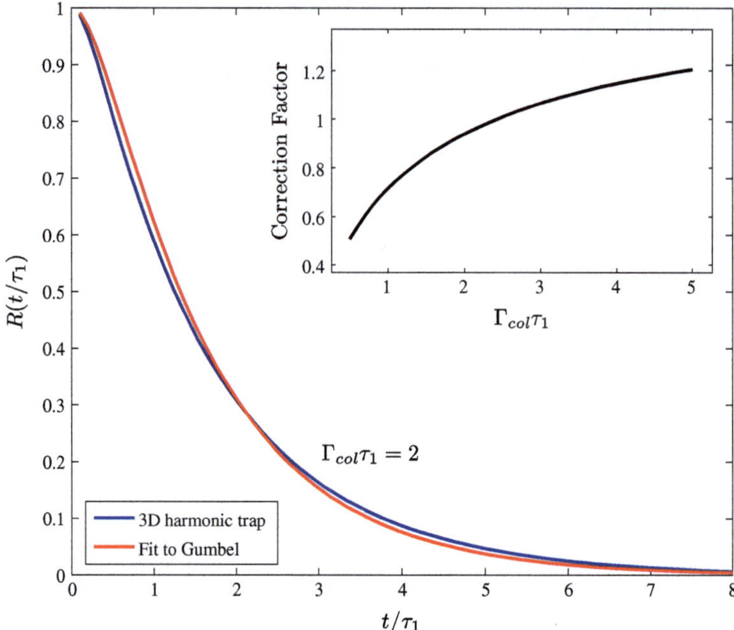

Fig. A.1 Comparison between a correct Ramsey decay envelope for 3D harmonic trap (*Blue*) and the fitted Gumbel function (*Red*) for $\Gamma_{col}\tau_1 = 2$. The inset shows the correction factor for the fitted value of τ_2 as a function $\Gamma_{col}\tau_1$

Appendix B
Derivation of the Spectrum for a Discrete Fluctuation Model

We consider an ensemble of TLS with fluctuations in the detuning which are discrete in nature and follow Poisson statistics. The probability for a specific TLS to undergo n events up to a time t is given by $\mathcal{P}_n = \frac{e^{-\Gamma t}(\Gamma t)^n}{n!}$, where Γ is the detuning randomization rate. The total Ramsey signal, $R(t)$, can be constructed as a series

$$R(t) = \sum_{n=0}^{\infty} \mathcal{P}_n(t) R_n(t), \tag{B.1}$$

where $R_n(t)$ is the Ramsey signal of the subgroup of TLS that experienced *exactly* n randomization events. The average over this group of any function $\hat{X}(t_1, ..., t_n)$, where $\{t_1, ..., t_n\}$ is a series of the n Poisson events times, can be written as

$$X(t) = \frac{n!}{t^n} \int_0^{t_2} dt_1 ... \int_0^{t} dt_n \hat{X}(t_1, ..., t_n).$$

We use this with $\hat{R}_n = \prod_{l=1}^{n+1} \int_{-\infty}^{\infty} d\delta_l P_0(\delta_l) e^{i\delta_l(t_l - t_{l-1})}$ and Eq. (B.1) in the paper to calculate $R_n(t)$

$$R_n(t) = \frac{n!}{t^n} \prod_{k=1}^{n} \int_0^{t_{k+1}} dt_k R_0(t_{k+1} - t_k), \tag{B.2}$$

where $t_{n+1} = t$ and $t_0 = 0$. Taking the Laplace transform of Eq. (B.2) and employing the n-fold convolution theorem we get $\mathcal{L}\{\frac{t^n R_n(t)}{n!}\} = \tilde{R}_0^{n+1}(s)$, where $\mathcal{L}\{f(t)\} \equiv \tilde{f}(s)$ is the Laplace transform of f. Substituting this into the Laplace transform of Eq. (B.1) we obtain $\mathcal{L}\{R(t)e^{\Gamma t}\}(s) = \sum_{n=0}^{\infty} \Gamma^n \tilde{R}_0^{n+1}(s)$, which can be further simplified:

Y. Sagi, *Collisional Narrowing and Dynamical Decoupling in a Dense Ensemble of Cold Atoms*, Springer Theses, DOI: 10.1007/978-3-642-29605-5, © Springer-Verlag Berlin Heidelberg 2012

$$\tilde{R}(s) = \frac{\tilde{R}_0(s+\Gamma)}{1 - \Gamma\tilde{R}_0(s+\Gamma)}. \tag{B.3}$$

Equation (5.1) gives the ensemble spectrum, $\tilde{R}(s)$, in terms of the bare spectrum $\tilde{R}_0(s)$ and the detuning randomization rate Γ.

Appendix C: Numerical Simulations with Correlations in the Fluctuations

Our aim is to generate an ensemble of a time-series of detunings $\delta(t)$ with some arbitrary correlation function $\langle \delta(t)\delta(t') \rangle = C(t - t')$, where $\langle \cdot \rangle$ stands for the ensemble average. We first explain how to generate a matrix of correlated numbers Y with a correlation matrix C from a matrix of uncorrelated numbers X. We assume that the numbers in X were drawn from the required detuning distribution. We look for a matrix U which satisfy $U^T U = C$, which can be found by using the Cholesky decomposition method. If we then define $Y = XU$ we get the desired correlated matrix, which can be verified by writing $Y^T Y = U^T X^T X U = C$, where we have used the relation for the uncorrelated matrix $X^T X = I$. In the simulation we start by drawing uncorrelated detunings from a Gaussian distribution. For each simulated atom we then draw a series of collision times $\{t_i\}$ from an exponential distribution with a rate Γ, where the index i stands for the i-th collision. We generate the correlation matrix in the following way: $C_{ij} = C(t_i - t_j)$. Finally, we use the matrix C to generate the correlated detunings series, and use it to calculate the contribution of this atom to $R(t)$. We then repeat this procedure for all atoms.

Y. Sagi, *Collisional Narrowing and Dynamical Decoupling in a Dense Ensemble of Cold Atoms*, Springer Theses, DOI: 10.1007/978-3-642-29605-5,
© Springer-Verlag Berlin Heidelberg 2012

Appendix D
Optimal Decoupling Pulse Sequence
for a Lorentzian Spectral Function

We want to find the pulse sequence timing series $\{t_i\}_{i=1}^{n}$ which maximizes the coherence as calculated by Eq. (3.5). We use a nonlinear minimization method (Nelder–Mead) implemented in MATLAB software for the minimum decoherence. We have consistently obtained a time series close to the form given in the text: $t_i = \frac{\eta+i-1}{n-1+2\eta}t$, where $i = 1 \ldots n$ and $0.5 \leq \eta \leq 1$ is a numerical factor which depends on the parameters of the problem. For $f_{DD} > \Gamma$, the difference of the time series for the optimal η to the one with $\eta = 0.5$ (which corresponds to the CPMG scheme) is less then 1%. As an example we plot the coherence time as a function of η for $f_{DD} = 35\text{Hz}$ in Fig. D.1a. The difference in this case between τ_c of $\eta = 0.5$ and the optimal value is $\sim 0.14\%$. In Fig. D.1b we plot the same calculation for $f_{DD} = 10$ Hz. Here the difference between the optimal value and the value at $\eta = 0.5$ is $\sim 0.93\%$. Therefore the value $\eta = 0.5$ was chosen to be used in the experiment.

In regard to the Uhrig dynamical decoupling scheme (UDD) [2], we plot in Fig. D.2 a comparison between CPMG and UDD as a function of the number of π-pulses. As can be clearly seen from the graph, the UDD scheme is inferior to the simpler CPMG scheme for a Lorentzian bath spectrum. This is not surprising since the UDD is expected to be superior in spectral functions that exhibit a cutoff in frequency. This point was already discussed in Ref. [3]. We have also tested this experimentally for several number of pulses and consistently obtained coherence times with the UDD scheme which are 10–20 % lower than the coherence times obtained with the CPMG scheme.

Y. Sagi, *Collisional Narrowing and Dynamical Decoupling in a Dense Ensemble of Cold Atoms*, Springer Theses, DOI: 10.1007/978-3-642-29605-5,
© Springer-Verlag Berlin Heidelberg 2012

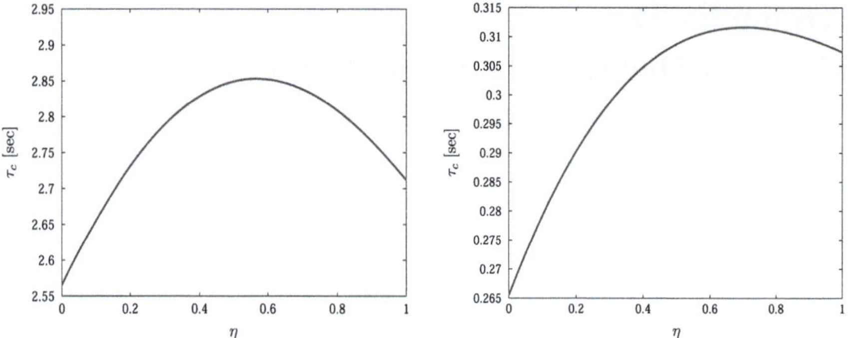

Fig. D.1 Coherence time versus η for $f_{DD} = 35$ Hz **a** $f_{DD} = 10$ Hz **b** with a lorentzian power spectrum with the typical experimental parameters $\Gamma = 37.5s^{-1}$ and $\sigma_\delta = 23.8s^{-1}$. The calculation of τ_c is done at $t = 1s$

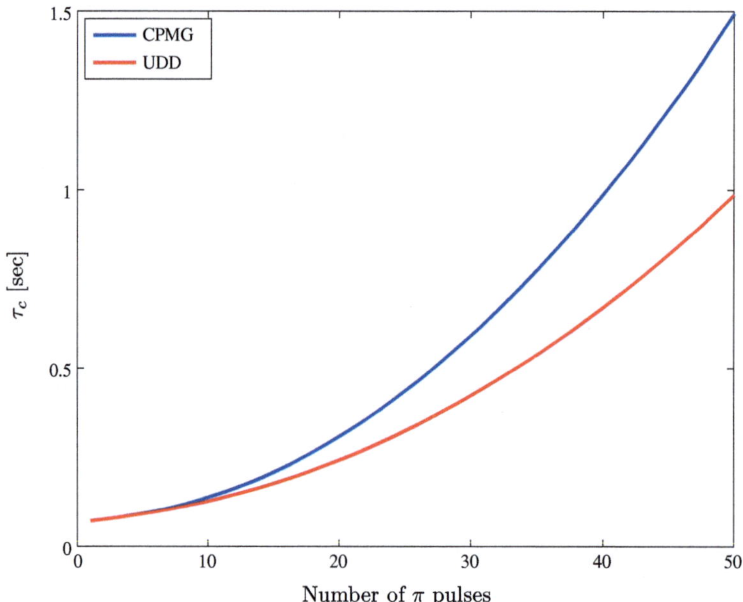

Fig. D.2 Coherence time versus the number of π-pulses for a CPMG dynamical decoupling scheme and the UDD scheme, for a lorentzian power spectrum with the typical experimental parameters $\Gamma = 37.5s^{-1}$ and $\sigma_\delta = 23.8s^{-1}$. The calculation of τ_c is done at $t = 1s$

Appendix E
Publications Resulting from This Research

1. Sagi, Y., Almog, I. & Davidson, N. Universal scaling of collisional spectral narrowing in an ensemble of cold atoms. *Phys. Rev. Lett.* **105**, 093001 (2010).
2. Sagi, Y., Almog, I. & Davidson, N. Process tomography of dynamical decoupling in a dense optically trapped atomic ensemble. *Phys. Rev. Lett.* **105**, 053201 (2010).
3. Sagi, Y., Pugatch, R., Almog, I. & Davidson, N. Spectrum of two-level systems with discrete frequency fluctuations. *Phys. Rev. Lett.* **104**, 253003 (2010).
4. Sagi, Y., Pugatch, R., Almog, I., Aizenman, M. & Davidson, N. Motional Broadening in Ensembles With Heavy-Tail Frequency Distribution. *Phys. Rev. A.* **83**, 043821 (2011).

References

1. Brissaud A, Frisch U (1974) J Math Phys 15:524
2. Uhrig GS (2007) Phys Rev Lett 98:100504
3. Cywiński L, Lutchyn RM, Nave CP, Sarma SD (2008) Phys RevB 77:17450

Y. Sagi, *Collisional Narrowing and Dynamical Decoupling in a Dense Ensemble of Cold Atoms*, Springer Theses, DOI: 10.1007/978-3-642-29605-5,
© Springer-Verlag Berlin Heidelberg 2012